Degrowth Decolonization and Development

Milica Kočović De Santo ·
Stéphanie Eileen Domptail
Editors

Degrowth Decolonization and Development

When Culture Meets the Environment

Springer

Editors
Milica Kočović De Santo
Institute of Economic Sciences
Faculty of Dramatic Arts
Belgrade, Serbia

Stéphanie Eileen Domptail
Institute for Agricultural Policy and Market Research
University of Giessen
Giessen, Germany

ISBN 978-3-031-25944-9 ISBN 978-3-031-25945-6 (eBook)
https://doi.org/10.1007/978-3-031-25945-6

This Springer imprint is published by the registered company Springer Nature Switzerland AG
The registered company address is: Gewerbestrasse 11, 6330 Cham, Switzerland

Contents

Chapter 1
How Culture and Worldviews Shape Development and our Environment

Milica Kočović De Santo, Stéphanie Eileen Domptail, and Jennifer Hirsch

Abstract Degrowth Decolonization and Development offers a collection of seven original case study analyses, followed by a synopsis of concepts contributing to decolonize development by shaking the hegemony of the Western paradigm. The participating researchers met when presenting their work in Decolonization and Degrowth panels within two International Degrowth Conferences held in July (organized by Manchester University together with the Ecological Economics conference) and in August 2021 in The Hague, Netherlands. Ranging from cultural studies, critical development studies, cultural policy, cultural political economy, political economy, heterodox economic approaches, eco-feminist political ecology, to anthropology and sociology, the collection of chapters provides a broad interdisciplinary oversight of the contemporary (developmental, environmental, economic, social, cultural) challenges. Precisely this interdisciplinary approach facilitates the understanding of the critical contemporary context with its complex intermingling of (positionality) crises. Our multiple analyses of Western thought, capitalist and patriarchal systems rooted in case studies depict rebellions to this hegemonic system and challenge it from complementary angles, which is the reason why we found the degrowth as most suitable framework to understand the current contemporary context and seek for post-growth alternatives. The first chapter introduces the most important concepts such as: permanent crisis, modernity and colonnialism, with associated dichotomies encaptured in the Western paradigm. We embrace the position that colonialism is not derivative but constitutive of modernity as "there is no modernity without colonialism" (Mignolo and Walsh 2018: 4; 107) and modern capitalism, where the cultural potential for radically necessary changes is essential—as a driver of the degrowth forces that reflects the eternal relations between man and nature.

M. K. De Santo (✉)
Institute of Economic Sciences, Faculty of Dramatic Arts, Belgrade, Serbia
e-mail: milica.kocovic@ien.bg.ac.rs

S. E. Domptail
Institute for Agricultural Policy and Market Research, University of Giessen, Giessen, Germany

J. Hirsch
Institute for Socio-ecological Research, Hamburger Allee 45, 60486 Frankfurt am Main, Germany

M. K. De Santo and S. E. Domptail (eds.), *Degrowth Decolonization and Development*,
https://doi.org/10.1007/978-3-031-25945-6_1

1.1 World Crises, Western Paradigm, Modernity and Colonialism

The list of current (state of permanent) crises and global challenges is overwhelmingly long: the climate crisis, species extinction, flood disasters, wildfires, poverty and famine—to name only a few. According to Moore (2015), these crises are neither convergent nor multiple, but singular and manifold. Capra maintained already in 1984 that the world's crises are unified through a "crisis of perception". This crisis in turn stems from the fact that we, as a Western *modern* society, apply an actually outdated worldview to a reality that has outgrown the concepts contained therein (Moore 2015: 3–4; Capra 1984: 15–16).

The concept of *modernity* as it is used in *Degrowth Decolonization and Development*, refers to a rational and science-driven perception of the world that entails a dualist framing—which Moore refers to as the Cartesian binary (Moore 2015: 21). Modernity is further characterized by the beliefs in progress, growth, and secularism (Mignolo and Walsh 2018: 106 et seqq.), beliefs that characterize the Cartesian Western paradigm, including the Scientific Method and originate in the Europe of the 17th century. Modernism is therefore more than a an entity or an ontological historical period; it is "a set of self-serving narratives [...] a construction made by actors, institutions, and languages that benefit those who built the imaginary and sustain it, through knowledge and war, military and financial means" (Mignolo and Walsh 2018: 110). In practice, modernity is tightly associated with the auspicious narratives of salvation, progress and development.

The darker side of Western modernity is revealed by the concept of colonialism. We embrace the position that colonialism is not derivative but constitutive of modernity as "there is no modernity without colonialism" (Mignolo and Walsh 2018: 4; 107). The concept of colonialism was first introduced by Aníbal Quijano in the late 1980s. Yet, it has a history and praxis of more than 500 years for it was established "by a selected community of humans of a given religion (Christianity), in a continent called Europe and around the fifteenth century, in the process of defining themselves as humans" (Mignolo and Walsh 2018: 6; 16; 153). As a matrix of power, colonialism today touches upon all aspects of life. Colonialism relations were explained as "domination and submission which turn the colonizing man into a classroom monitor, an army sergeant, a prison guard, a slave driver, and the indigenous man into an instrument of production (...) colonized societies drained of their essence, cultures trampled underfoot, institutions undermined, lands confiscated, religions smashed, magnificent artistic creations destroyed, extraordinary possibilities—wiped out" (Cesaire 2000) colonialism brought the destruction of anti-capitalist indigenous cultures and economies.

Its three pillars i.e. racism, sexism and last but not least the invention of the concept of Nature (Mignolo and Walsh 2018: 10, 159) constructed the colonial differences. Its *modus operandi* involves the destruction and denial of worldviews, knowledges, beings, spirituality, lands, ways of thought and (cosmo-)existence, social cohesion and everything else, that threatens to impede its way (Mignolo and Walsh 2018: 15–16).

Considering the further worsening global challenges of the 21st century, it seems appropriate to assume that Western imperial states are no longer in control of the monster (i.e. the colonial matrix of power) they created and that "the universalized model or paradigm of the West is crumbling before our very eyes" (Mignolo and Walsh 2018: 10; 15). Previous patterns of action are no longer effective in solving systemic crises. These patterns have mainly supported human technocratic omnipotence's by developing ever more innovations that have made it possible to exploit nature (including human nature) in yet more powerful ways and to enjoy its "free gifts" on the cheap. The failure of those strategies that perpetuated capitalism over the last 500 years now indicates a pending change of epoch (Moore 2015: 1). With colonialism, capitalism arrived at such a pivotal point (Mignolo and Walsh 2018: 16).

The decolonization discussed here is not solely directly related to settler historical colonialism (see critic of Tuck and Yang 2012); rather it denounces the hegemony of a particular worldview originated in Europe among scientists and intellectuals, and the theories and disciplines that were built during the last centuries in this Western European context. The exportation of this line of thought through economic and development theories, their practices and institutions (e.g. World Bank, International Monetary Fund etc.) to culturally and geographically different areas is the form of colonialism the book challenges.

Decolonization thinking and doing intend to disengage from the epistemic assumptions valid within all areas of knowledge of the Western world since the 16th century. This goes beyond offering resistance against the West's global model and rather aims at *re-existence*, which refers to the endeavor to redefine and reorient the praxis of living in conditions of dignity (Mignolo and Walsh 2018: 3; 106, accentuation by the author).

However, decolonization and a decolonial pluriversal perspectives does not negate Western thought *per se.; it* understands it as integral (cf. Litfin 2003; Moore 2015). A decolonial pluriverse perspective shall rather require an ongoing serpentine movement toward possibilities of other modes of being, thinking, knowing, sensing, and living; that is, an otherwise in plural" (Mignolo and Walsh 2018: 3; 81). Such a perspective is particularly important for attempts to decolonize economics and the governance of our environment and societies as it helps to reduce the risk of viewing decoloniality from the *outside* of the colonial matrix of power only. Looking at decoloniality from the outside would not only diminish the spheres of action, but would "also blind eyesight of the decolonial cracks that exist within this matrix and system and that, in essence, complement and push toward the edges and borders" (Mignolo and Walsh 2018: 82). These cracks and fissures within the Western Christian civilization as "spaces, places, and possibilities of and for decolonization" (ibid.) are the major focus of interest in this chapter and in *Degrowth Decolonization and Development*. In fact, *Degrowth, Decolonization and Development* address these contemporary systemic crises of energetic supply, mining, extractivism, risk management, corporate food systems, climate change and gendered thinking, among others.

This first chapter introduces our investigation of the role of culture and its transformative potential in the management of the current global crises by viewing these as a single crisis of perception, by reflection on historical findings form self-governance in Yugoslavia as endogenous knowledge from semi-periphery and the dissonant heritage as the alternative social and economic organization that introduced the social ownership. It successfully demonstrates how the social, the economic and the ecological spheres are inseparably linked to the cultural sphere. In each case study, our cultural/historical approach reveals the historical, institutional and discursive interrelations that drive the management of our earth resources and thereby provides a cultural understanding for the emergence of these crises. It points to specific efforts on the ground but also alternative concepts dismantling the hegemonic paradigm and offering spaces to meander towards another perception of humanity in the world and the future.

1.2 Interdisciplinarity and Positionality in Cultural Analyses

The book *Degrowth Decolonization and Development* addresses issues, topics and challenges from complementary theoretical and practical spheres relevant for the management of our global crises. Our common premise recognizes culture as a *man-made environment* is an essential sphere and leverage point for the management of societies (Meadows 1999): it holds transformative and progressive potentials, because it is the sphere of values and meanings that defines the goals of societies within the life system space.

Ranging from cultural studies, critical development studies, cultural policy, cultural political economy, political economy, heterodox economic approaches, eco-feminist political ecology, to anthropology and sociology, the collection of chapters provides a broad interdisciplinary oversight of the contemporary (developmental, environmental, economic, social, cultural) challenges. Precisely this interdisciplinary approach facilitates the understanding of the critical contemporary context with its complex intermingling of crises. Our multiple analyses of Western thought, capitalist and patriarchal systems rooted in case studies depict rebellions to this hegemonic system and challenge it from complementary angles, which is the reason why we found the degrowth as most suitable framework to understand the current contemporary context and seek for post-growth alternatives.

Indeed, each chapter of the book addresses a specific case, in which a minority or marginalized group rebels against actions, institutions or values rooted in the Western worldview. Methodologically, the case studies unfold to analyze institutions, formal or informal, and values or beliefs, guiding development and in particular the interaction of humans with(in) nature. The analyses and the issuing recommendations presented in each chapter feed into a deconstruction of the Western paradigm through a criticism of its constitutive dichotomies. They also reveal principles which will feed into the design of novel systemic solutions making use of "pluriverse" alternatives rooted in diverse cultural backgrounds, as a decolonization process per se. This structure fosters the interdisciplinary integration of knowledge for the common epilogue.

Thus, our approach generates insights instrumental in achieving a theoretical, political and practical decolonization, as well as Degrowth imaginaries, beyond the concept of sustainable development within the growth paradigm. Our imaginaries and trans-disciplinary conclusions featured in the last chapter seek to accompany readers on integrative paths leading to environmental justice, social and economic equity and equality.

As editors and authors, we have given particular attention to our own *inner eyes* which means that our research magnitudes and inspirations are also inherited through our personal cultural capital building, what the decolonial feminist thinker Sylvia Wynter (2003) called the classificatory logic. It frames and restricts how we see, know, and act on (Mignolo and Walsh 2018: 17, our accentuation). The consequential dilemmas of our writings in our various roles and positions as authors have to be recognized. Some of us are white privileged women from Western Europe attempting to write about and reflect on the worldview of a minority or culture(s) they do not belong to. The dilemma of partially reproducing the dominant modern structure while we attempt to investigate other worldviews has to be accepted for its insolvability (cf. Mignolo and Walsh 2018: 20–21). Some of us, women from Central Europe and China, stem from the very countries they investigate in their research, and attempt to write critically from within, despite potential risks. They attempt to create a new perspective while intrinsically raised and marked by the very logic they seek to criticize.

The third case is that of Adam Pomieciński, who is investigating a case of gold mine extraction in rural Armenia, and the role of social movements in the Western Asia with the eyes of a Polish man researcher. Both countries arguably belong to the periphery of the global North, albeit in varying extents and self-perceptions. In all cases, such a cultural analysis and critique is contingent on our initial positions, values and cultures. This exercise demands a constant reflection of our own socialization and relation with Western thought. Our intention is to think *from and with* (not only *about*) subjects, which "advance other ways of being, thinking, knowing, feeling, and living [...] that interrupt, transgress, and fissure or crack modernity/ coloniality's matrices of power, and make evident concrete instances and possibilities of the otherwise" (Mignolo and Walsh 2018: 20).

By acknowledging the relativity of knowledge and the existence of parallel ontologies, we also support the idea that decoloniality should not be considered an individual attribute, or a lineal point of arrival or enlightenment. Instead, we seek to "make visible, open up, and advance radically distinct perspectives and positionalities that displace Western rationality as the only framework and possibility of existence, analysis, and thought" (Mignolo and Walsh 2018: 17). This is also the objective of this *Degrowth Decolonization and Development.*

1.3 Central Dichotomies Regarding the Western Paradigm, Modernity and Coloniality

A major conceptual pillar in our cultural analyses is the focus on conceptual dichotomies characterizing the Western dominant worldview, such as Global North versus Global South; Center versus Periphery; Developed versus Developing/Underdeveloped; Man versus Nature. All chapters involve specific cases of countries or populations standing in a relation of historical hegemonic disadvantage (Poland, Serbia, Armenia, Japan, China, indigenous populations, agro-ecological farmers, women (...)) and facing major institutional change and processes rooted in modernity and leading to extractivism, democratization or organization.

Starting from Marxist studies, via the Frankfurt School to reach the present, in the shape of feminism, ecology and post-colonialism amongst other liberatory pulses, critical theory is, in essence, concerned with the critique of modernity (Munck 2016). The dichotomies opposing Global South to global North; Developed to developing; Center to periphery, are conceptually close and explained by critical development theory (CDT) (Munck 2016). A*s an alternative to t*his dichotomous classification lens (Wynter 2003), CDT offers a *«post-development» imaginary which unmasks the whole development enterprise as an illusion which left no viable challengers to its orthodox development theory* (Munck 2016). CDT refers to several authors in terms of fundamental issues and concerns: *the collapse of globalism* (Saul 2005); or the liberation of *human beings from the circumstances that enslave them* (Horkheimer 182: 244). Many more that built the base for CDT *explain what is going wrong with the current social order, identify the agents for social change and provide practical goals for social transformation*—a transformation that enables the *culture turn* towards a «post-development" era (Munck 2016). Also, the essential contribution is given by the degrowth (francophone, Spanish and Anglo-Saxon) schools of thought, that are mostly involved in the challenges of growth paradigm.

Central to the modern/Western/Newtonian/Cartesian worldview and thus directly implicated in the vast violence, injustice, and oppression of modernity in its core, is the binary of nature and society. This central dualism, man-nature, enables the view of nature as external, quantifiable and rationalizable. The detachment of humans from nature is not only a crucial condition for capital accumulation (Moore 2015; Merchant 2020). It is also intricately entwined with modernity's knowledge structures (i.e. the logic of contradiction and the semantic construction of binary opposition (Mignolo and Walsh 2018: 154–155)), its prevailing power relations, its status of re/production, prosperity, as well as modern human behavior patterns towards their environment (Moore 2015: 2–3). Hence, what the world faces today "is not a crisis of capitalism *and* nature but of modernity-*in*-nature. That modernity is a capitalist world-ecology." (Moore 2015: 4, our accentuation).

Furthermore, if we tend to name the things, we must understand that capitalism and colonialism are from the same river stream. We shall no separate these two concepts. From its history of slave trade, draining the culture, cheap labour and nature etc.—were the real reasons for colonial conquests. Capitalism relies on appropriation of cheap labour and nature that enables profit to grow, its character is *colonial by nature*. As such it seeks for universalism, not the pluriversalism. This means that culture is deeply connected in processes of justifying contemporary ideologies and systemic injustices and inequalities that includes racial and gender issues. It is a systemic order of global inequalities, through which all existing dichotomies and unbalanced accumulation of wealth can be easily understood.

Departing from this capitalist world-ecology culture will require overcoming the separation of man and nature. The power relations characterizing this separation must be redefined, both in the relation between people and nature and as a result also between Man and woman, and Man and minorities.

1.4 Cultural Change for a Decolonized Development and Human Relation With(in) Nature: A Collection of Case Studies

Degrowth Decolonization and Development offers a collection of seven original case study analyses, followed by a synopsis of concepts contributing to decolonize development by shaking the hegemony of the Western paradigm (Table 1.1). The participating researchers met when presenting their work in *Decolonization and Degrowth* panels within two International Degrowth Conferences held in July (organized by Manchester University together with the Ecological Economics conference) and in August 2021 in The Hague, Netherlands.

Milica Kočović De Santo's *The Progressivity and Transformative role of culture—Findings from self-governance in Yugoslavia towards life-centered development* present a historical analysis of the cultural policy as integral part of total social and economic development in Yugoslavia between 1953 and 1990. She herself was born in Yugoslavia and has witnessed both the socialist and the capitalist economic system as well as the separation of the country in six nations, form which some were formed as national states for the first time in their history. Cultural policy in Yugoslavia was always deeply interlinked with economic policy, politics, ideology, the state, and (the concept of) development. By identifying what in culture is financed by the state and how, Kočović De Santo says that it is possible to read the ideology in every term (*ideology apparatus* and the *repressive apparatus)*. The analysis smoothly depicts the linkages between economic theory, culture, development, critical cultural studies, critical development studies, and degrowth by bringing the reflection beyond the concept of sustainable development and growth paradigm, towards the post-growth

Table 1.1 Overview of the case studies investigated in *Degrowth Decolonization and Development*

Authors	Book chapter	Case study	Crises addressed	Dichotomies addressed	Theoretical contribution
Milica Kočović De Santo	Chapter 2 *The Progressivity and Transformative role of culture—Findings from self-governance in Yugoslavia towards life-centered development*	A historical analysis of the cultural policy with special focus on the self-governance management socialism in Yugoslavia	Capitalism and its hegemony, crisis of well-being, equality and justice	Developed versus developing, semi-peripheral knowledge versus Western knowledge on cultural policy and governance	Life-centered Development based on inherited endogenous knowledge—as dissonant heritage of self-governance in Yugoslavia and other relevant theories
Inga Kuźma	Chapter 3 *Anticipation of the Degrowth Concept in the Socialist Republic of Poland of the 1970s—The Case of the Artistic and Scientific Movement*	The case of early pro-environmental cultural events in Poland in the 1960s	The global environmental crisis and the global hierarchy among countries	East versus West, developed versus developing, peripheral, post-dependence	The works of eco-art social activism, including feminism and anti-discriminatory efforts, as a bridge for future visions of a post-anthropocentric world
Adam Pomieciński	Chapter 4 *Radical movement against global growth—The gold mine Amulsar in Armenia*	Protests around the gold mining project of the Amulsar Firm in Armenia	Inequalities, environmental protection and rural livelihoods and identities	Global North versus global South, center versus periphery, capitalism versus livelihoods	Development of anti-capitalist logic and a revision of the Degrowth concept from the middle-east rural perspective

(continued)

Table 1.1 (continued)

Authors	Book chapter	Case study	Crises addressed	Dichotomies addressed	Theoretical contribution
Katharina Richter	Chapter 5 *Framing Cultural Directions of Degrowth Transitions: Cosmological Limits to Growth and Affective Abundance* contributes to degrowth debates *on limits to growth*	A study of indigenous worldviews in Ecuador, Latin America	The reliance of capitalism and growth in the economy to sustain livelihoods and the Earth system	Growth versus non-growth Global North versus Global South Human versus Nature Science versus non-science	An alternative to development and scarcity, thanks to decolonial theory and practice from Latin America; reciprocity; Limits to growth
Stéphanie Eileen Domptail, Jennifer Hirsch, Ernst-August Nuppenau	Chapter 6 *Decolonizing nature? Worldviews of agroecological farmers in Germany to address the global environmental crisis*	Worldview of agroecology farmers (members of the German La Via Campesina movement) in Hesse, Germany	The food system crisis with its productivist paradigm, leading to environmental damages while currently unable to feed well the world population	Human versus Nature Science versus non-science Mind versus matter Production versus reproduction	Decolonizing human-nature relations in food production system, with theoretical borrowings from Feminist economics, environmental sociology and political ecology. Reproduction and life basis; Co-creation with nature Spirituality
Tamara Schneider	Chapter 7 *Aestheticizing Catastrophes? A Comparison of the Western and Japanese Approach to Art Creation in View of the Climate Emergency*	A comparison of two poems reflecting on major climate catastrophes stemming from Japan and the USA	The global environmental crisis, our attempts to harness climate change, power relations with Nature	Human versus nature Anthropocentrism to Ecocentrism	The critical art sociology reflects on the human relation with(in) nature and thereby demonstrates the potential transformative role of culture in economic development and climate policy debates

(continued)

Table 1.1 (continued)

Authors	Book chapter	Case study	Crises addressed	Dichotomies addressed	Theoretical contribution
Ting Wang	Chapter 8 *The absence of gendered management in climate change in China: an analysis of adaptation policies from a gender mainstreaming perspective*	China's policy texts, designed to mitigate climate change effects and foster climate adaptation	The limits of the patriarchal, rational, objective governance system	Man versus women Science versus non-science	Gender mainstreaming as a mean to enter the policy-making arena is a less disruptive manner, and as first step to enable the internal challenge of the patriarchal policy making system

futuristic idea of *life-centered development*. The aim of this new concept is to integrate equally cultural and environmental issues pertaining to specific endogenous contexts of life in the heart of future economic policies. Kočović De Santo questions the roots of the permanent crisis, which she locates in the contemporary economic logic where the dichotomic world in permanent crisis seeks for the urgent radical and progressive changes. Contrary to the vast of majority of Degrowth scholars, Kočović De Santo argues that the needed change should stem from the Global South and so-called Periphery countries, who bare a historical hegemonic disadvantage. Countries of the Global South can choose pluriverse paths and real alternatives as exit strategies from the global wreckage, instead of applying exogenous knowledge as a "*proven paths*", Kočović De Santo seeks to create possibilities for the integration of valuable endogenous practices and knowledge, inherited from non-capitalist systems outside the Global North such as the workers self-governance in Yugoslavia developed after WWII into future policy making. This would contribute to the trans-disciplinary integration towards the further elaboration of the concept of life-centered development. The life-centered development is driven by a sphere of peoples' "*loud voices*" and *progressive transformative force, who,* besides culture, equally share environmental values, conviviality, good life and common sense. Its "*silent voices*" of nature transcend anthropocentrism in dealing with environmental challenges. By spreading virtue values from human-human, towards human-nature, such as solidarity, empathy, equity, justice, empowered by eco-feminist vistas, life-centered development becomes an essential base for decolonized future degrowth economies and policies.

In her contribution *Anticipation of the Degrowth Concept in the Socialist Republic of Poland of the 1970s- The Case of the Artistic and Scientific Movement*, Inga Kuźma also performs a historical analysis of Degrowth using interdisciplinary anthropological cultural studies as a frame. As a polish female anthropologist, Inga Kuźma reports on the case of early pro-environmental cultural events in the Poland of the 1970's, thus embedded in the geopolitical field of Central-Eastern Europe, at the time of formation of the Polish socialist system. The art and music events analysed here constitute, as Kuźma argues, the case of Polish degrowth in its beginnings through eco-artivism. Kuźma presents the degrowth idea as a social and historical fact, i.e. the alternative project of society growth which should be treated as universal in spite of past geopolitical obstacles, such as the "Iron Curtain" between the West and the East. This period witnesses the blossoming of eco-art and feminism interests (contemporary to one of the first appeals of the UN, warning against a degradation of Nature attributed to the development of human civilization (a. o. a nuclear arms race, industrialization, urbanization, progress of consumption). Against the geopolitical local background inspiring that event similar to those depicted by Kočović De Santo, Kuźma brushes the cultural picture of so called 'peripheral' economies, with their politics and cultural imagination of civilization, advance and improvement in the Polish socialist past system (which follow the Western capitalistic values). She shows that the discourses of care, against environmental exploitation and the critique towards consumption and overproduction constituted a common and parallel strain of resistance between the West and the East, between the Center and the Peripheries. Kuźma's interdisciplinary approach as a dialogue between art—activism—ecology—science and humanities is best able to reveal the extent of the impingement of Degrowth at the time. The crucial conclusion is that Degrowth in its present shape was founded on the complex interplay of diverse reflections conducted through art and research. In particular, art has (had) the freedom to act subversively in anticipating the future and exploring new models of solution. The works of eco-art social activism, including feminism and anti-discriminatory efforts, represent the bridge for the future visions of post-anthropocentric world.

In his chapter *Radical movement against global growth—The gold mine Amulsar in Armenia,* Adam Pomieciński critically reflects on extractivist mining projects, by presenting and analyzing the case of the Amulsar gold in Armenia. Pomieciński is a polish cultural anthropologist who has conducted several research work in Western Asia. Here he reports on the controversy among both the inhabitants of Armenia and environmentalists that emanated from the investment of the Amulsar Firm into the mining project. Activists and independent experts pointed to the risks associated with the construction of a gold mine, including irreversible damage to the natural environment, destruction of the entire water system of Armenia and destruction of the cultural heritage of the region. Despite these arguments, the construction of the gold mine was completed in 2018, and mining began. As a reaction, the indigenous environmental movement took radical steps to block activities in the Amulsar area and protests continue to this day. The example of the Amulsar gold mine in Armenia shows the consequences of global growth for the peripheral countries. The agents of contestation are as often activists and grassroots movements, recognized in Kočović De Santo (this book) as the *allies of change* particularly important as it appeared through similar interventions regarding the comparative project of Rio Tinto mining investigation of lithium in Serbia, that she investigates in her researches. Pomieciński reflects on the radical activities and actions of the environmental movement related to the exploitation of gold in Amulsar, and shows explicitly how these are tightly linked to the global critique of growth goals in the context of anti-capitalist logic through the degrowth points.

In her chapter, Katharina Richter's *Framing Cultural Directions of Degrowth Transitions: Cosmological Limits to Growth and Affective Abundance* contributes to degrowth debates *on limits to growth*, political ontology and abundance. Richter is a female German scientist based on the UK who has conducted her extensive study of indigenous worldviews in Ecuador, Latin America. Her analysis of the Kichwa *Buen Vivir/sumak kawsay* (BV/sk) worldview presents an alternative to development from Ecuador, enabling her to put degrowth into conversation with decolonial theory and practice from Latin America. By establishing this inter-epistemic dialogue, she contributes to efforts of provincializing degrowth by involving Kichwa cosmological views on *the limits of growth.* This notion complements recent degrowth, ecological economics, and political ecology debates that juxtapose external, physical boundaries with morally constructed, internal limits to growth. Cosmological limits to growth are normative constraints to the destruction of the living world. 'Cosmological limits to growth' recognizes that while 'limits' are both socially constructed and physically present, their relational understanding places limits on growth before the breach of socio-ecological ones. Richter gives impetus to discussions around cultural direction of degrowth processes beyond socioeconomic transitions. The collective production of *psychological well-being through relational justice* offers lessons to the cultural politics of degrowth. It embeds nature into the social world not just from a natural capital or ecosystem services perspective, but a cultural and political one.

The sixth chapter *Decolonizing nature? Worldviews of agroecological farmers in Germany to address the global environmental crisis* focuses on the role of mental models in designing sustainable food systems based on a renewed relation of humans with(in) nature. In their chapter Stéphanie Eileen Domptail and Jennifer Hirsch supported by Ernst-August Nuppenau explore the worldview of farmers belonging to a famous agro-ecological peasant movement in central Germany. Both researchers and researchers are rooted in Western European culture and embossed by its values; yet both researcher and researchers seek to create/identify cracks in this construction leading to alternatives. Thereby, Domptail and al. address two philosophical problems with the productionist paradigm. First, productionism focuses on production and ignores the role of supporting social and environmental processes coined reproduction processes in feminist economics. Second, productionism is nested in the Western worldview, in which Nature is understood through a colonized, utilitarian and largely extractivist lens and Humans are separated from Nature, a key dichotomy in the Western paradigm. It thereby ignores the systemic and emotional *rapport* to nature of agricultural production. Domptail et al. show that German farmers attempt to decolonize their relation to Nature and within the food system through their activities. Mental models shared among the agro-ecological farmers appear in part radically opposed to the utilitarian extractivist worldview. Domptail et al. describes in detail how human-nature unity, spirituality and mystical experiences, reproducing the life-basis and egalitarian relations to animals and other humans in the food system play a key role in the worldview of the agro-ecological farmers. Their analysis points towards the existence of a holistic and egalitarian,—nearly organic—worldview among agroecology farmers, which is highly modern as farmers consider explicitly the impact of their actions on the global social and ecological world in their decisions. The case of the agro-ecological farmers shows admirably the iterative interactions through which practices and culture co-evolve.

In her chapter *Aestheticizing Catastrophes? A Comparison of the Western and Japanese Approach to Art Creation in View of the Climate Emergency* Tamara Schneider uses critical art sociology to compare poems reflecting the human relation with(in) nature and stemming from Japan and the USA as a reaction to extreme climate catastrophes. She herself is German and based in Japan. Schneider picks up the transformative role of culture highlighted by Kočović De Santo and Kuzma by pointing out how closely interwoven culture is with society and environment. Furthermore, she connects to the debate on limits to growth addressed by Pomieciński and by Richter, by discussing Western and Japanese art pieces. The selected contemporary artworks show different approaches to present "natural" disasters. Schneider argues that these different approaches act, on one hand, as evidence of divergent concepts of our environment. On the other hand, these examples also reveal that Western concepts and perception patterns are often considered to be universally valid and applicable, even in the case of aesthetics. However, looking at the artworks, different philosophical worldviews shine through, which largely differ from this universalism. Critically questioning established Western concepts facilitate the decolonization of the traditional concept of economic development. In the current climate policy debate, mutual cultural understanding could, e.g., make global climate negotiations more successful and provide much needed diversity in climate mitigation and adaptation. The latter also immediately connects with crisis management broached in the farming analysis by Domptail et al.

The eighth chapter *The absence of gendered management in climate change in China: an analysis of adaptation policies from a gender mainstreaming perspective* finally deepens the feminist perspective and its critic of the patriarchal system characterizing the" Western" paradigm. As a young female researcher who comes from China and now studies in Japan, Ting Wang explores her chapter based on her long personal and academic experience in living and studying in China and attempts to provide a critical analysis of China's climate policy design from within. Rooted in China's culture, she seeks to identify potential gender perspective in China's climate change, using a content analysis of climate adaptation policies. Wang provides one of the rare gender-aware analysis of the topic of climate change research in China. She shows that climate change adaptation policies remain gender-blind. At the same time, she reveals that adaptation strategies in China mostly rely on the typically "masculine" domains of technology from an energy and technical perspective rather than reflecting on the role of society and people, from different genders. These attributes depict that policies remain designed in a resolutely patriarchal worldview/perspective. Wang adopts the gender mainstreaming perspective as a mechanism to improve women's political participation and to obtain equal voices in processes of effective responses to climate change, as a first step towards achieving gender equality and women's economic and political empowerment. She emphasizes that achieving gender mainstreaming is not the final purpose: rather it must be seen as only a first step to in a second step fundamentally question and transform the grounds of this patriarchal power structure, where pathways towards solutions are not limited to spheres of technocrats. Responding to the climate crisis, she claims, requires us to challenge existing inequalities, reshuffle our values as well as the structures current policymakers safeguard that organize our society.

Finally, our synopsis *Beyond Western dichotomies of power: egalitarian power relations, affective abundance and co-creation within nature* synthesizes the conclusions of the respective case studies with regard to the cracks and fissures in the Western paradigm that were or currently are being generated in the cases reported upon. These cracks take the form of dismantling the binary classification lens of the modern to decolonize the hegemonic worldview. Posing the question of how to overcome these central dichotomies is for us an instrument to relativize the Europe-centered world perspective that dominates in Global North imaginaries, to break the power relations in the controlling and dominance in our relation with(in) nature, to reveal dark sides of patriarchal systems. Rather we open doors to new ways of knowing and acting which shall stimulate the development of new imaginaries and soft models for systemic solutions paving the way towards the radical change needed. These include possible Degrowth alternatives that integrate decolonized knowledge, co-creation with nature, changing societal goals to life-centered development, to the maintenance of the life basis, systems thinking, cultivating and using affective abundance.

In the world and even within Europe and the global North, alternative worldviews, maybe suppressed ones, are (re-)emerging, calling to be heard, as part or inspiration for a solution to global crises. Giving attention to these suppressed voices is the decolonization in sense of practice the book contributes to.

Degrowth Decolonization and Development is structured as a collection of seven original case study analyses followed by a synopsis on decolonizing concepts. The participating researchers met when presenting their work in *Decolonization and Degrowth* panels within two International Degrowth Conferences held in July (Manchester together with the Ecological Economics conference) and August in Hague 2021.

Degrowth Decolonization and Development provides a critique of concepts associated with the term "Development", which will certainly be instrumental in the decolonization of both development theories and practices. This critique is made possible by exposing knowledge and theories outside the comfort zone of the mainstream Western theoretical landscape. The book convincingly reveals the large array of domains, which, when interpreted from a decolonization and Degrowth perspective, can be managed through fundamentally different logics and generating societally more desirable outcomes. The book thereby reveals common underlying cultural roots to a diverse array of crises.

This meta-level cultural perspective will widen the perspective of all those who seek to march beyond their disciplinary knowledge and paradigm. From a geographical cultural perspective, both readers from the Global North and the Transition Countries will identify to the cases presented. Readers from the Global South will witness how concepts stemming in part from authors of the Global South are used to deconstruct a dominant paradigm originating in the Global North. *Degrowth Decolonization and Development* is an accessible introductory text to decolonization in Development studies as it relates to several important disciplines, such as climate and food systems, culture, epistemology, economy, and anthropology.

References

Capra F (1984) The turning point. In: Science, society, and the rising culture. Bantam Books, New York

Cesaire A (2000) Discourse on colonialism, a poetics of anticolonialism. Monthly Review Press, Printed in Canada

Litfin K (2003) Towards an integral perspective on world politics: secularism, sovereignty and the challenge of global ecology. J Int Stud 32(1):29–56

Meadows D (1999) Leverage points. Places to interviene in a system. The Sustainability Institute, Mariland. https://donellameadows.org/wp-content/userfiles/Leverage_Points.pdf

Merchant C (2020) The death of nature: women, ecology, and the scientific revolution. Harper Collins, Digital New York

Mignolo WD, Walsh CE (2018) On decoloniality: concepts analytics praxis. Duke University Press, Durham/London

Moore JW (2015) Capitalism in the web of life: ecology and the accumulation of capital. Verso, London/New York

Munck R (2016) Critical Development Theory: results and prospects. https://www.researchgate.net/publication/315729676_Critical_Development_Theory_results_and_prospects

Saul JR (2005) The collapse of globalism and the reinvention of the world. Viking, Camberwell

Tuck E, Yang KW (2012) Decolonization is not a metaphor. Decolonization 1(1). https://jps.library.utoronto.ca/index.php/des/article/view/18630

Wynter S (2003) Unsettling the coloniality of being/power/truth/freedom: towards the human, after man, its overrepresentation—an argument. CR New Centennial Rev 3(3):257–337

Milica Kočović De Santo is an interdisciplinary senior researcher at the Institute of Economic Sciences in Belgrade, Serbia. She investigates alternative systemic solutions in the quest for needed radical transitions through social theories, involving economic, developmental, decolonial, commons, and cultural studies that meets in the field of degrowth. She teaches at the Faculty of Dramatic Arts and University of Arts in Belgrade, Serbia.

Stéphanie Eileen Domptail is a senior researcher at the professorship for agricultural, food and environmental economics at the Giessen University (JLU), Germany. Stéphanie applies concepts of ecological economics and feminist economics to the study of agricultural and food systems. Stéphanie investigates agroecological systems in Germany, Nigeria and Brazil. She brings together the topics of ethics, decolonization and agroecology in their joint capacity to redesign human-nature relationships towards a more desirable state.

Jennifer Hirsch works at the institute's executive board in the department of scientific services at ISOE- Institute for Social-Ecological Research in Frankfurt, Germany. She studied American Studies and Business Administration at Goethe University Frankfurt and Nutrition Economics at Justus Liebig University Giessen. In her master's thesis at the University of Giessen, she investigated the conception of nature and the worldview of agroecological farmers in Germany.

Chapter 2
The Progressivity and Transformative Role of Culture

Findings from Self-Governance in Yugoslavia Towards Life-Centred Development

Milica Kočović De Santo

Abstract Until the 60s, in Yugoslavia, the emphasis was mainly on the economic policy and instruments—in a narrower sense. Traditionally unprofitable activities started to appear to capture in a more excellent picture of development. The culture slowly became *an integral part of social being* by allowing the politics and policy to become social constructs (through self-governance socialism) and not the exclusively the construct of political elites. *Cultural development represented one of the few most critical dimensions of social development, integrated in* the sense of total development. Historical analysis regarding the cultural policy in sustainable development, and vice versa, will provide a better understanding of the place of the culture and its importance, for example, in Yugoslavia and Serbia. The critical discourse and content analysis, scoring on the endogenous knowledge, culturally-driven and further more environmentally-driven factors, will help capture the contributions of culture to a greater or lesser extent to the paradigm of sustainability in history of Serbia and its Yugoslav heritage. Historical conclusions are fundamental grounding for future assistance in re-solving environmental, social, cultural and economic issues and challenges resulting from economic policy trends and pressures. Economic history is not frozen, instead it is still being written, highly depended on the present: What is the relationship between culture, development, and sustainability from decolonized lenses? How and why it is still relevant to apply commonly inherited aspects of knowledge regarding Yugoslav self-governance in the contemporary context? What is meant by life-centered development—to be done for the future? Why is it essential to start form the decolonization of knowledge and epistemic erasures towards imagining future integrative cultural and environmental policies?

Keywords Yugoslav self-governance · Sub-periphery cultural policy knowledge · Critical development studies · Decolonization · Eco-feminist economics · Political ecology · Degrowth · Life-centred development

M. K. De Santo (✉)
Institute of Economic Sciences, Faculty of Dramatic Arts, Belgrade, Serbia
e-mail: milica.kocovic@ien.bg.ac.rs

M. K. De Santo and S. E. Domptail (eds.), *Degrowth Decolonization and Development*,
https://doi.org/10.1007/978-3-031-25945-6_2

2.1 Introduction

The chapter's primary goal is to analyze the institutional changes as interlinked processes (of culture, economy, and society) by focusing on the field of culture throughout the history of Serbia and Yugoslavia. Findings will enable us to determine the role of culture and cultural policy in developmental trends. The power, politics, and ideology will be presented in the paper through the mutual relationships among the culture, the economy, and the social development.

Findings from different historical phases (from the first, the second and third Yugoslavia until today (Đukić 2010; Dragičević Šešić 2018) will provide a basis for understanding changes by stages and influencing factors (such as ideology, identity, and politics); the work of Western academic hegemony in terms of invisibility of theoretical and practical knowledge regarding the theoretic development of Yugoslav (as semi-periphery and Global South) praxis thought of cultural policy and political economy. Also, Yugoslav direct contributions to authentic "third ways"—through internal social organization of the Yugoslav self-governance socialism followed by introduction of social ownership in free market; and Yugoslav decolonization legacy in launching the Non-Alignment Movement are still intriguing pieces of history.

The analysis involves dichotomies of state versus free market, centralization versus decentralization, Global North versus Global South, Centre versus Periphery, in the manner of knowledge creation, relevance and factual importance as international decolonization legacy.

In various historical phases, the cultural policy, more or less, explicitly contained a specific ideology. The main findings from different historical phases will reflect institutional changes throughout history until today's transitional cultural policy that shares attributes with modern economic policy—to enable understanding its "permanent crisis" context. This chapter will provoke to provide an universal endogenous (dissonant) grounding, for modeling alternative systemic solutions for future fairer development, based on economic and cultural policy.

2.1.1 The Main Points of This Chapter

The contextualization and application of theoretical findings interlinked with historical phases of culture and cultural policy in Yugoslavia will provide a broader context and highlight the contemporary importance of tracing the place of culture (Hawkes 2004) in development. The main question is how to capture the contemporary cultural context due to interdepartmental relations from different historical phases influenced by the broader economic sphere, politics, system, ideology and governance? What (and where) is the power of culture to make development fairer? Answers to these questions, except from the mentioned invisible or forgotten semi-periphery authors form Yugoslavia, will be mainly supported by the cultural political economy (CPE) approaches (Sum and Jessop 2013; Jessop 2016; Jessop and Sum 2006; Albittron et al. 2007).

Furthermore, the discussion will include questions on what we have learned and created in terms of endogenous knowledge? How the hegemony exogenous vs endogenous looks like today, due the work of *"academic capitalism"?*

Concerning the answers to these questions, basic theoretic grounding comes from critical developmental studies (CDS) (Escobar 1992; Munck 2016; Castro 2004; Sachs 1992; Rist 2010), decolonization and post-colonial theories, with special focus on epistemic erasures.

The start point of research is to understand the position(ality) of Serbia as a country in hegemonic disadvantage (Global South, Periphery, and *country in the race to the bottom*), where the endogenous knowledge often dissonant heritage, is the cultural basis for thinking about the decolonization of knowledge and development in parallel directions (Kočović De Santo 2021a, b, c, d; Latouche 2009; Mbembe 2015; Hudson 2020; Kapoor 2008; Dragičević Šešić 2018) as beneficial approach to possible future progressive changes led by culture within degrowth—as a synthetic theory with a strong decolonization, critical and green movements legacy. This part of the research will include critical approaches in terms of development by summarizing a critical theoretical overview, by leading us to the second point of understanding challenges and seeking solutions by involving different sources except from the mainstream and comfort zones of knowledge, which eventually translates to the work of degrowth vistas in practice seen as the decolonization of knowledge.

Understanding the challenges based on historical analysis will shed more light on the relations among culture, economy, society and environment, including ideology, politics, and power, by enabling the discussion through decolonized lenses on the paradigm of sustainability toward the meaning of equitable, fair and desirable development, and the ways to achieve it in the future.

Again, this leads us to interdisciplinary crossings between different theoretical approaches from CPE, CDS, and CCS to the eco-feminist (Bauhardt and Harcourt 2019; Berik and Kongar 2021; Dengler and Seebacher 2019; Dengler and Strunk 2018; Gaard 2017; Mellor 2015) heterodox economic, evolutionary economics and sustainable transition market theories (Christoforou and Laine 2014; Petri 2021; Wray 2020; Arestis and Sawyer 2019; Köhler et al. 2019; Boon et al. 2020; Rip and Kemp 1998).

By synthesizing the previous, it will be possible to propose the modeling of new systemic solutions for alternative futures by placing the life (culture, society more integrative with environmental context) in the heart of development by re-thinking the options out of the consumerist sense - as a culturally driven process. Here Hawkes's insights on the interdependency of sustainable development, flourishing culture and the fundamental role of culture (Hawkes 2004), as well as part of imagining and shaping public policies (Hawkes 2013), are crucial for further inferring the meaning and sense of (good) life on a personal and societal level.

2.1.2 Methodology and Concepts

This chapter relies on critical discourse, where the analysis is structured as text–context—interpretation, including historical analysis until today on self-governance with the special focus on culture and cultural policy knowledge of the semi-periphery. The synthesis through inter-disciplinary research, respecting the interdependence of theories, policies, politics, ideology, and governance, will lead to trans-disciplinary discussions and conclusions on the "permission to say self-governance"—as dissonant heritage and knowledge from semi-periphery, that will pave the ways on the role and power of culture beyond sustainable development towards desired degrowth transitions.

The starting research point is about involving Hawkes views (2004) that I am taking as a primary grounding in terms of arguments for the central hypothesis.

H: Culture holds the progressive and transformative forces for future systemic changes. It is a fundamental reason to involve the findings form workers self-governance in Yugoslavia, especially in terms of its cultural aspects: *as dissonant heritage and endogenous knowledge from sub-periphery* for the future degrowth model of fairer development for the good life.

Society's values are the basis upon which everything else is built. The cultural sphere encompasses values and the ways they are articulated and expressed. Furthermore, integrating the values into the governance models is about the ***culture at work*** (Hawkes 2004).

Governance without expressing community values cannot be fully democratic (Hawkes 2004). Achieving the governance framework as a social construct means integrating the culture in policymaking processes based on more "democratic and participative" democracy. I consider historical lessons as reflections from Yugoslav workers' self-governance as endogenous knowledge and dissonant heritage (form the semi-periphery and Global South) to be incorporated in modelling radically needed change through the decolonisation of knowledge and development in theoretic and practical terms.

Enable active community participation, Culture in Action as the beginning of a Great cultural challenge (Hawkes 2004). By this point, I will go beyond Hawkes—towards the beginning of Great cultural change! Development and the economy are part of people's culture; There is much more life, than the economy can explain (Hawkes 2004). Including only financial aspects of life as contemporary macroeconomic indicators (GDP, GDP per capita) inspires the sense of life in the *Catch-22* consumption-production matrix; towards articulation and recognition of the pure purpose and relation and feedback loop of materialistic other non-materialistic processes and aspirations of social reproduction.

H1: Bringing the **life** into the heart of development is the main goal for the future. Involving care work and other processes that contribute to the quality of life without which a system could not work, means the result of degrowth praxis, led by culture (values, meanings) involving rights of (and for) nature. Proving this hypothesis will be supported by the theory of cultural political economy (CPE), evolutionary economics and sustainable market transitions. That will allow a clear discussion with historical analysis on culture and cultural policy and its role in the system of self-governance development (in Yugoslavia).

Discussing the Marxist theory of the state, besides the state as the repressive apparatus,[1] Althusser distinguished the ideological state apparatus, recognising a certain number of realities which present themselves to the immediate observer in the form of distinct and specialised institutions,[2] concluding that what matters is how institutions functions, by minimising the sector (private or public) (Althusser 1971). If culture is understood as the factory of values, meanings, and social relations, then the link between state and culture shall be a permanently interrelated feedback process, not one-sided.

The cumulative genesis of the modern state given by Willke (Jessop 2016) is a suitable framework for understanding the evolution and challenges of the state, politics, system, and development paradigm. By spreading genesis towards Althusser's *ideology state apparatus,* the interrelated fields regarding the culture, society, and environment become clearer. As part of the cumulative genesis of state study, cultural relations, sociology, political economy, and informatics, the evolution in education allowed scientific support for creating practical policies. The purpose of cultural policy is to define the humanistic functions of culture based on academic settings, which follow the philosophy and general development tendencies conditioned by the ideology of a specific society. Cultural policy *is the practical policy of the state administration bodies and other bodies holding political, legal and financial authority to make important legislative and executive decisions in the field of culture*; its content consists of studying the models, instruments and strategies used by these bodies that are formally in charge for cultural policy (Đukić 2010).

Cultural policy gained special significance after WWII when many states started implementing the cultural policy as a public policy. After war, states were encouraged by the *UN Universal Declaration of Human Rights.*[3] Dichotomous divisions characterised the global after-war period. The *Iron curtain* erected a barrier (territorial, ideological, political, economic and cultural). The central division of the time was East versus West (Blocks). In the Eastern Block, left-leaning countries introduced various socialist economic and market development approaches, while the Western

---

[1] (1) The state is the repressive state apparatus, (2) state power and state apparatus must be distinguished, (3) the objective of the class struggle concerns state power, and in consequence the use of the state apparatus by the classes (or alliance of classes or of fractions of classes) holding state power as a function of their class objectives, and (4) the proletariat must seize state power in order to destroy the existing bourgeois state apparatus and, in a first phase, replace it with a quite different, proletarian, state apparatus, then in later phases set in motion a radical process, that of the destruction of the state (the end of state power, the end of every state apparatus).

[2] The religious ISA (the system of the different churches), the educational ISA (the system of the different public and private 'schools'), the family ISA, the legal ISA, the political ISA (the political system, including the different parties), the trade-union ISA, the communications ISA (press, radio and television, etc.), the cultural ISA (literature, the arts, sports, etc.).

[3] *The International Convention on Human Rights* and various other international human rights documents, and, in particular, by the provision on the right to participate freely in the cultural life (Article 27).

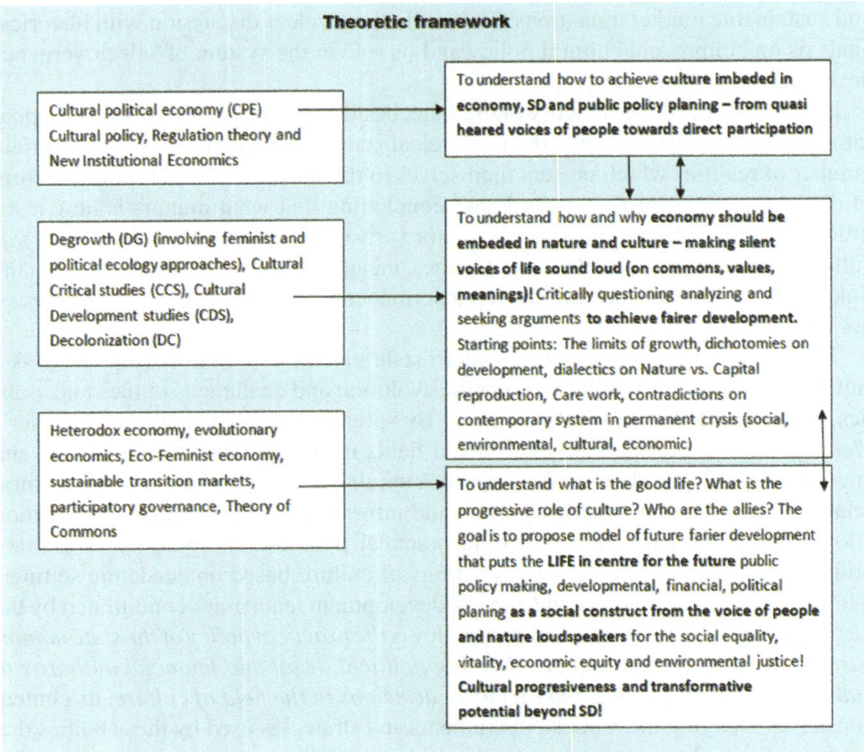

Fig. 2.1 Theoretic framework

Block openly promoted a liberal-capitalist society and the free market economy. At the same time, international financial assistance organisations were established in both blocks.[4]

The appearance of Global North and Global South is firmly tied with Truman's speech,[5] where he introduced *development as a concept* followed by interpretations of *underdeveloped and developing countries*. The development and its meaning have brought dichotomies ever since. Namely, it implied the division of the world into *developed, developing, and underdeveloped; Global North versus Global South, Centre versus Periphery*. Underdeveloped countries from different backgrounds[6] had been interpreted exclusively in terms of economic growth (Fig. 2.1).

[4] The Organisation for Economic Co-operation and Development (OECD) was established involving the members of the Western bloc. In economic terms the counterpart for OECD within the Eastern Bloc was The Council for Mutual Economic Assistance (CEA).

[5] In 1949.

[6] Geographical, cultural, economic, level wealth considering real economy, natural and cultural goods and resources.

2.2 Critical Developmental and Decolonization Discourses—Towards Deeper Understanding of the Progressive and Transformational Potential of the Culture

The critics of development included broad topics dependent on the historical momentum, from social, economic and later environmental and eco-feminist issues of unfair distribution, basic needs, poverty, education, environmental sustainability, governance, and access to education; sanitation, health, drinking water, and other fundamental human rights. Most of these issues are related to exogenous constraints to development given as a "*universal proven path*" given by developed, for others to follow. Endogenous knowledge of "others" was not considered as relevant due to (*epistemic erasures*) which become more relevant today, as *pluriverse* concept of indigenous cosmologies. The main point of universal development proposal is that it excludes many endogenous contexts, such as specific cultural, social or environmental dynamics. The issue of development as a theory and practice to overcome poverty and inequality has not gone away as a vital global concern (Munck 2016). The question today is whether and how theories and practices can meet to overcome these challenges? Munck explains the evolution of critical development studies and theory CDS as a theoretical stream from primarily critical studies (CS) spring; Critique is the essence of dealing with the modernity issues (Munck 2016). Munck offers theoretic evolution from CS, Frankfurt school, social theory, towards CDT to explain "*what is wrong with the current social order and who are the agents for social change and provides practical goals for social transformation, adding that we need to find a new theory of imperialism if CDS is to find life and applications in practice*" (Munck 2016).

Furthermore, culture and cultural critique are recognised as an essential part of the poststructuralist views. Moreover, "*new forms of cultural critique must emerge in the spaces of negotiation among an increasing number of detailed spheres of expertise and interest*" (Marcus and Fischer 1996; Pieterse 1996).

Critical developmental insights relevant for the place of culture are given mainly through post-Marxist approaches on relation power-knowledge in terms of critique of Eurocentric and mainstream discourses (Munck 2016). The developmental paradigm brings the imbalances, instead of solving them (Kapoor 2008); inequalities, exploitation by unifying cultural diversity, towards cultural homogenisation in the function of same systemic order capitalism[7] while constructing an object of knowledge and intervene to extend state power by objectifying people and nature in the Third World (Castro 2004; Rist 2010; Kothari et al. 2018); where the role of social movements in building up alternatives of social and cultural change is recognized as essential (Escobar 1992). From the view of Escobar, "*development was conceived not as*

[7] The cultures of the periphery need to be replaced by utilitarianism, which is the culture of capitalism.

a cultural process[8] *but instead as a system of more or less universally applicable technical interventions intended to deliver some "badly needed"* goods to a *"target"* population"; While pointing that economic ideas are not universal truths, by offering a historical analytical path to show how they penetrate and constitute the *"economist culture of modernity"* (Castro 2004). At the beginning of the 80s, Worsley indicated that *"culture was the missing concept in development studies, when a few years after the United Nations proclaimed the World Decade for Cultural Development, putting culture and development firmly on the conceptual and policymaking terrain"* until the end of the 80 s it becomes clear that economic development *"cannot be dissociated from the cultural context in which it is embedded"* (Munck 2016).

These are practical adverse outcomes, as opposed to pluralism and democratic tendencies. Although modernisation theory is usually referred to as a paradigm, it turns out to be the host to a wide variety of projects along the lines of endogenous change,[9] at the same time, it has also been associated with projects of exogenous change[10] (Pieterse 1996) leading often to the undesired outcomes that can be explained through the *incommensurability term* that refers to have no common to measure, due to incompatible value points.

The first point of the chapter is about answering whether the contemporary cultural context as a result of interdepartmental relations from different historical phases is influenced by the broader economic sphere, politics, system, ideology, state and governance; and the power of culture to make development fairer? I start from the previous relevant theoretical findings that will be intersected with historical phases of culture and cultural policy in Yugoslavia to highlight the *place of culture* [2] in (sustainable) development, with the special focus on the self-governance phase. Answering the questions assumes encompassing complex interrelations.[11] It will introduce the concepts of the state and its institutional changes, supported by cultural political economy CPE and historical analysis of different stages (Yugoslavia until now), to bring the specific context regarding the role of the culture in total development.

The next step is to involve critical developmental and cultural studies overviews to capture the inter-linkages around culture. The cultural policies will be presented within different historical phases and sub-phases, with particular attention on the second phase of Yugoslavia and workers' self-governance, as inherited endogenous knowledge and an essential history lesson—regarding the socio-economic organization and the role of culture—relevant for the future degrowth development.

The final part of the discussion is related to the trans-disciplinary integration of knowledge aiming to reflect decolonised arguments for future life-centred development.

---

[8] *Culture was a residual variable, to disappear with the advance of modernisation.*

[9] Social differentiation, rationalisation, the spread of universalism, achievement and specificity;

[10] The spread of capitalism, industrialisation through technological diffusion, westernisation, nation-building (nationalism as a derivative discourse), state formation (as in postcolonial inheritorstates).

[11] On politics, policy, governance, culture, SD, ideology, economy, system, etc.

2.3 The Permission to Say "Workers Self-Governance in Socialist Yugoslavia" as a Dissonant Heritage

In terms of values, the origins of freedom, equality and brotherhood came from eighteenth century as liberal-democratic heritage, were almost naturally extended and radicalized towards all society members through the international socialist movements in nineteenth and twentieth centuries. The capitalist economy of that time was not static was expanding relatively quickly, where the impersonalized free market provided successful organization—where no one could be blamed for upcoming miseries. So there was no reason to expect the revolution as a product of development as such, but only as outcome due unbearable social tensions. The gradual accumulation of wealth through the free market, although unfair, was tied at the level of all classes, which loosened social conflicts, along with the political rights that disadvantaged classes received. Even organizing into labor unions and parties actually contributed to the reduction of social tensions. On the other hand this was the potential to involve new priorities of labour movements seen through the stronger political and economic participation, and power re-allocations. Marxist and Post-Marxist theories were indispensable as starting point that provided analytical frameworks for the study of polarization and inequalities in society, based on class analysis to identify social antagonisms in a simplified way, by understanding minority, majority, marginalized, racial groups, etc., whose conflicts are the basis on which capitalism rests. But the conflicts are also the basis from which capitalism deviations can be resolved or changed. French revolution appeared as a negation of feudalism, while Russian revolution brought the negation of private ownership. Both revolutions contributed to the better economic organization and efficiency.

To meet the stronger participation, opposite then private or state capitalism, self-governance appeared as alternative model of socialism to be achieved within non-authoritarian regime and society. Up to 1951. Yugoslavia was classified (by UN and other international bodies), along with the Soviet Union and the other East European countries, as having a planned economy. After the 1951. Yugoslav economy was excluded from the "Soviet bloc" or the "socialist camp" (Horvat 1969). Horvat claimed that initial stages of Yugoslavia's independent development of socialism, were determined geographically by international audience, and assigned to its development (as a country of Western or Southern Europe). In practice there were two pure types: the private capitalist type with the so-called free market, represented, for example, by the United States, and the centralized planned type with state ownership of the means of production, represented, for example, by the Soviet Union, Along with these, there were other, intermediate forms, accentuating one or the other component more or less, as is the case with the "welfare economies" of some countries of the West Europe where Social Democratic parties are in power (Horvat 1969). Yugoslav economy has become an authentic project. Obviously, it was not a private capitalist, because private ownership of the means of production has been abolished. Yugoslavia also abolished centralized planning in less than 6 years (from the end of IIWW), by introducing comprehensive institutional changes, based on the idea of decentralization of power and authority. This brought the autonomy to social and economic agents, by supporting the framework for their direct decision-making processes. The Yugoslav economic system of self-governance was based on a single fundamental principle—*social ownership*—of the means of production and capital, which was the source of its unity, as a type of economy that Horvat described as *associationist socialism* (Horvat 1969). Here the questions of authority, power allocation and the enterprise played different roles than in central planned economy.[12]

The economic democratization in Yugoslavia was carried on the process of horizontal and vertical decentralization. To explain the previous, I will originally cite whole Horvat paragraph that brings the comparisons: *The Yugoslav solution, like the Swiss one, belongs to the horizontal and vertical decentralization. However, there was a huge difference between the two, which in this context comes down to the fact that a Yugoslav commune practiced much more economic intervention than a Swiss canton, and that, on the other hand, Yugoslav producers practice much more direct political intervention than those who control Swiss economic organizations (…) Yugoslav experience has given us historical evidence: worker self-governance is not only possible in practice (it has been functioning in Yugoslavia for over a decade) but is economically efficient (since it was restored, the Yugoslav economy has reached the highest rate of growth in the world)* (Horvat 1969).

In 1970. Federal Executive Council, made a request due to the first stoppage in the mid-60 s, a group of the most relevant experts (who were representatives of the republics of the Federal Republic of Yugoslavia) was hired[13] to conduct the research in two weeks. The group was formed with the aim of creating the so-called *White Paper on "Economic Functions of the Federation"*. The outcome of the requested research was based on critical analysis, statistical data and scientific integrity, freedom and objectivity of internationally known Yugoslav experts of that time. The idea was to propose economic policy instruments and solutions that will ensure the harmonization of a Federal market as single and unique. Negative economic trends appeared as an outcome due the inadequate set of Federation functions (burdened with administrative bureaucratic mentality and insufficient social responsibility), as well as ongoing improvised and insufficiently professional economic policy (Horvat et al. 1970). The group of experts came to the challenges in terms of ongoing dichotomies, which are addressed in the following relations:

[12] The authority is directly relevant to the problem of centralization and decentralization A nationalized branch of industry that is run by the state may be formally organized on the same principles as a system of worker self-governance. "Centralism here—centralism there". But there was an essential difference The first case involves a political authority that dictates; the second involves a social authority to which the workers' collectives submit voluntarily because they realize their economic and social interests (Horvat 1969).

[13] The group consisted of prof. Hasan Hadžiomerović University of Sarajevo, prof. Dr. Branko Horvat, scientific advisor of the Institute of Economic Sciences in Belgrade, prof. Dr. Nikola Kijušev, University of Skopje, prof. Dr. Branislav Šoškić, director of the Institute for Economic Research in Belgrade, prof. Dr. Dragomir Vojnić, scientific advisor of the Economic Institute in Zagreb.

(1) *Centralization or Decentralization*? In the Yugoslav economy, decision-making was maximally decentralized on work organizations, while the Federation should have intervened minimally only when necessary;
(2) *The state apparatus as a government or a Social service*? The state apparatus is first and foremost the apparatus of government. In this sense, the state apparatus focused almost exclusively on *granting permits and enforcing prohibitions*. Issuing direct orders was no longer possible due to the obvious consequences for the main economic entities whose state of affairs influence was *little or not at all*. Because of this, the efficiency of the entire system decreased and our society suffered serious losses. Thus, *the re-organization of the apparatus into a state service*, following the example of the Scandinavian countries, was suggested;
(3) *Federation understood as a State or a Yugoslav community*? The document offered an understanding of the *Federation as a Yugoslav community of people*, where economic decisions at the level of the Federation should have represented decisions for the Yugoslav economy as a whole, as a reflection of society. In this context, the expert group suggested reform in Federation system through 4 types of institutions in charge[14];
(4) *Authorization through the Federation or Agreements between the republics*? The results of the analysis indicated that "*the federal government should be given all the necessary powers, and the control of governments work should be carried out through the analysis of the results of their work. If the proclaimed program is not realized, a critical discussion regarding the trust that was given to the Government shall be open. If the deviations from the proclaimed goals, accepted plans and published predictions are significant, the government should resign. If we want to replace the mystified aureole of power in the state administration with the characteristics of public service, then the government should be treated to a large extent as the business board of this society*";
(5) *Socio-political communities (DPZ) or Economic organizations?* There is a clear labor social division between economic organizations and DPZ bodies. Business organizations gain income in the competitive struggle on the market. DPZ bodies harmonize the interests of economic entities in a certain territory. If the first were asked to harmonize the market, we would get a monopoly guided by rational choices, if we asked the others to enter the market; we would get etatism (Horvat et al. 1970).

Democracy in the period of Yugoslav self-governance was primarily manifested in the effort to include the most comprehensive strata of society in decision-making processes related to the most critical social issues. In this way, the effects of decision-making were in line with the systemic socialist goals of development.

The economic policy of the 1960s was about material production as a necessary precondition for the operation of the entire social mechanism. The development of

[14] The first is the supreme political (Assembly) body that develops economic policy on the basis of the broadest social consensus. The second is the Government, whose authority should be to set up by *expert-technical support* and *implementation of economic policy*. Third, expert-advisory bodies, and Fourth, interest associations such as trade unions and chambers (Horvat et al. 1970).

production power was determined by the quality and efficiency of the human factor. If we take into consideration that the culture was understood as one of the preconditions to form conscious and committed people of the humane goals of socialism; It is easy to read the socio-economic significance, progressive, transformative capacity and role that culture played for the society. The democratic self-governance mechanisms could ensure "*purposeful, rational decisions and optimal development of all social activities only if they were relying on competent and responsible subjects of political life*"; explaining further that the "*responsible subjects of political life can be created and reproduced only in a society in which culture occupies a high place*" (Madžar and Popov 1968) that will be more explained in next part. The development in workers self-governance Yugoslavia, was a totality of intertwined, connected, dynamic aspects of society as a process that is carried and oriented by society's values and needs (Kočović De Santo 2021a, b, c, d). The importance of consensus on social justice as a social aspect of development was the main focus in workers self-governance in Yugoslavia. Although culture does not necessarily have an explicit role in the total system, it has been an interdepartmental sphere connected with continuity since the 1960s to date.

The socialist dynamics of production in the Yugoslav worker's self-governance represented at first place a *new form of organisation of social life*, then secondly *economic life* in both terms—practical sense and qualitatively, was reflected in *the enormous developmental potential contained in socialist production relations can be fully realised only if the appropriate cultural development is ensured, by encompassing it on every human being* (Kočović De Santo 2021a, b, c, d), that is relevant as a contribution to the history of degrowth especially in terms of policy and financial mechanisms.

Financing and allocation for the *general*[15] *and common*[16] *needs* in self-governance is important to understand the financing of culture as it represented a common need. The system of financing general social and common needs included a set of financial measures, instruments, mechanisms, institutions, financial flows and social relations to support general and common consumption through socio-political communities (društveno političke zajednice DPZ) and self-governing interest communities (samoupravne interesne zajednice SIZ), as authentic financial associations in the self-governing socio-economic and financial system. The distribution of newly created values depended on four forms of consumption: personal, investment, general and common (Ristić 1982). Certainly, it is possible to say that objectivity was essential in the establishment of correlations between the mentioned types of consumption,

[15] It assumed the coverage of needs within budgetary and socio-political institutions, but also (in 60s and 70s) education and training, scientific activities, social care and protection, as well as communal activities, economic investments and investments in the economy. The financing system included two key subsystems, the tax (direct and indirect) and the budget system (Ristić 1982).

[16] According to the composition and priorities, income distribution was carried out according to primary: education, science, culture and health, and secondary: social protection, child protection, physical culture, employment and disability pension insurance. Also, part of the national income was transferred through contributions as support for non-economic activities, (Ristić 1982) as participatory contribution to the total development.

where the economic activities were the basis for the functioning of non-economic ones. In other words, general social and common consumption directly depended on personal income, social accumulation, personal income, personal consumption, turnovers of goods and services, etc.

In the system of financing general social and common needs, the acquisition of income was realized: (1) on a fiscal basis, (2) on the basis of freedom of work, (3) on the basis of providing services, (4) on a credit basis, and (5) on the basis of reciprocity and solidarity (Ristić 1982).[17] Sources in the system of financing general social and common needs were transformed and transferred through the self-governance allocation system. The financial flows represented money flows, as a reflection of the exchange between organizations of *joint work,*[18] *working people* on the one hand, and *socio-political communities,*[19] *self-governing interest communities*[20] of social activities, aiming to meet general and common societal needs in the overall flow of social reproduction. Workers in non-economic activities, such as the field of culture also contributed from their net income for common consumption funds and reserve funds to cover the functioning of common and general social needs (Ristić 1982).

Common needs and common consumption represented an essential part in the vision of building a new society by representing the sphere within which a balance was established, between material and non-material factors of socio-economic development. The activities of common consumption were aimed to support the development of a versatile socialist personality, while satisfying authentic needs by introducing elements of humanization into the sphere of consumption. Common needs were seen as essential by contributing to the formation of new civic identities, supported by libertarian values. Common consumption carried out the social equalization by mitigating differences in living standards, harmonizing regional differences, through meeting common needs, and gradually erasing differences in rural–urban needs, through the policy of unification. However, the share in participation to support and consumption of activities to satisfy the common needs—was uneven and directly related to the material base and other economic factors (Ristić 1982). This leads us to the conclusion that the access to the cultural, educational, health and science services was not exclusively the result of fundamental rights.

---

[17] Depending on the point of view, financial sources for social and common needs were the instruments of economic policy (fiscal) for the realization of economic and social functions of social reproduction. That is, financial sources at the same time represented an obligation regarding the calculation technique, but they were also the basis of earning income for DPZ and SIZ. Social and common needs were also serviced through social and bank loans, which in the end resulted in increasing costs of taxpayers, while in terms of the growth of social consumption, it was carried by the multiplier effect (Ristić 1982).

[18] Labour organizations, agricultural cooperatives, working communities of financial markets (insurance, banks, and other financial organizations).

[19] Municipalities, communities of municipalities, inter-municipal, regional communities, cities, provinces, republics, federations.

[20] Education, health, science, culture, social, child protection, employment, health, disability-pension insurance.

Social financing of common needs (social, cultural, scientific, educational, health development) assumed the provision of a material basis that was aligned to planning relations: goals-needs-priorities-value system-economic policy-resources-consumption-effects (Ibid.). The material base was provided (directly and indirectly) by horizontal connection of workers from economic activities with workers from non-economic activities of total social reproduction, during which the common needs financing mechanisms were established, and later supplemented by income relations. The common needs and the financial processes were carried out on the principles of: (1) decentralization and de-etatization, (2) democracy and self-governance, (3) pluralism of self-governing interests, (4) principle of financial balance, (5) principle of budgeting, (6) principles of financial coverage and financial federalism, (7) the principle of priority and selectivity, (8) the principle of financial coordination, harmonization, sovereignty, autonomy and co-financing, (9) the principles of solidarity and reciprocity, etc. (Ristić 1982).

According to Ristić (1982) at the end of the 80s, as he believed that it was still important task to achieve *the long-term process of creating approximately equal opportunities for all users of common services, for which the basic prerequisite is greater socio-economic equalization and the growth of the social well-being of working people and citizens*. Either way, the value-ideological elements of self-governing socialism, encompassed by legal-regulatory institutional changes, provided the basis for decentralized socio-economic and "more than human" development, where the income contributions from the economy were the source of contributions from the non-economy. While contributions as alimony for shared consumption, *represented at the same time a source of social income for certain social categories, which directly contributed to social equalization in workers' self-management in Yugoslavia* (Ristić 1982).

It becomes apparent, that achieving the decolonised development plan in workers self-governance Yugoslavia paved the ways towards decolonisation in many senses: decolonisation of knowledge—by seeking it out of the main-stream and comfort zones. Yugoslavia contributed to the decolonization processes by being directly involved in pioneering launchings of two authentic projects:

1. Yugoslav self-governance within market economy—as truly endogenous know-how interpretation of "how socialism shall work" in economic, organizational, regulatory and ideological manner. Project worked at states level for impressive time, having in mind external trajectories.
2. Yugoslavia carried out a valuable international policy and exchange with both the West and the East, but its greatest international contribution to decolonization reflected through the launching of the Non-Aligned Movement. By Non-aligned Movement a real alternative was created, by increasing visibility to the largest part (initiating decolonization of still colonized) of the world that did not fall into spaces of the dichotomous borders of the Iron Curtain. This enabled and pushed for-real the decolonisation of Global south countries, through defended independence of the unseen Third world.

2.3.1 *Why the Self-Governance is Essentially Important for the Future Degrowth Policies?*

Self-governance in Yugoslavia was authentic form of joint labor. It lasted four decades, constantly built, re-shaped and re-defined through the institutional changes. In comparison with the other economic systems of a given time, it is possible to distinguish specific features of self-governance in Yugoslavia.

(a) **Social ownership of the means of production**

Basic means of production (not all) were collectively owned, or more precise, social ownership was introduced in the self-governance. That implied the main distinction—in terms of self-governance organization processes which were completely different from public ownership. Most of other socialist central-planned economies held centralized political power. As such they represented a sort of *state capitalism*, seen through the lenses of the most influential praxis and self-governance researcher Horvat (1969). This means that such central planed models encouraged the hegemony of the (autocratic) political elites who etatized the management processes of public property. The property was the only essential distinction of central planed socialism in comparison to other capitalist forms, where in both cases the direct democracy was suppressed.

This was also the fundamental difference between the (lessons from) self-governance, contemporary capitalistic system, (where the means of production are mostly under private capital ownership) and state capitalism of centrally planned socialist economies (with the public ownership in the hands of elites and politicians). Also, the fact that decisions in self-governance depended on market conditions through the strong decentralization represented the main difference in comparison to the central planned economies in most of other socialist cases. It is usefully to mention that private ownership existed, just not in all departments and branches of total economy. For example, in both central planed administrative socialism phase (up to 1953), as well as during the workers self-government model (1953–1990) private ownership in means of production existed in agriculture, some industries (such as textile), crafts and tourism.[21]

(b) **Management of the means of production**

Another specific point of self-governance in Yugoslavia is the fact that management processes were highly democratic, decentralized and probably the „true last **direct democracy** from the Balkan fields ". The social ownership of means of production was delegated in terms of governance on the bodies such as *working collectives*

[21] Self-governance included re-construction of ownership that could be traced historically by the double-negation processes of ownership. At first place just after the IIWW the negation of private ownership took place—towards the public ownership followed by revolutionary processes of nationalization. We could say that Yugoslavia had only one "petoletka" five year of centrally planned economy. After 1950., the institutionally reforms appeared to enable the second ownership negation—form public—to social ownership.

and *working organizations*. These bodies were **autonomous**, and directly in charge to bring the decisions regarding relevant issues.[22] The decision-making processes were direct (included also secret voting) through the self-governance bodies (workers councils). Self-governance represented a resilient alternative, and a balance between a centrally planned and a modern capitalist economy of those times, as market oriented economic system that followed the economic laws.

By Šoškić self-governance was seen as **a corrective factor for market economy** (Šoškić 1982). He believed that it is not possible to replace economic laws, orientation, the way of market works, especially in terms of prices, by recognizing that in middle of 80 s that our economy needed radical turn in terms of theoretic and practical revitalization of self-governance in sense of building the society capacities to understand the importance of joint work. He detected "the main problem" of self-governance in Yugoslavia—as the slowing of economic growth, which is essentially valuable for the degrowth. Namely, self-governance structural "problems" lied in the fact that simple reproduction was possible in most of the cases, while the extended reproduction was the challenge. Also other issues regarding the income allocation, employment, inflation rate, price stabilization and international trade, were in the focus to "fix" (Šoškić 1982).

(c) **Amortization is the base to achieve simple reproduction**

By calculating amortization (depreciation) rate, in the society (that holds the role of the owner of total means of production), ensures the preservation of the value substance of those needed assets. However, many objective problems related to amortization—appeared. Besides the allocations for the simple reproduction, the global trend spreaded on the necessity for extended reproduction, to enable the material base of joint work as Marx mentioned in *Critique of Goth program*. With such changing course, many relevant researchers of that time perceived that private business and ownership of other (except basic) means of work, were far more favorable in the conditions of the so-called small economy (Šoškić 1982).

It can be concluded that the present forms of empowered participatory governance are most closely associated with workers' self-management that still exists in various forms of re-established cooperatives of joint work. The sectoral partnerships are more than desirable and ownership structure should strive to re-establish social ownership mechanisms in terms of shared resources, goods and services. As such, self-governance is unwritten history of degrowth, as it represented the systemic solution which was sensitive on societal values still relevant for the contemporary degrowth movement. Also, it could directly bring fair shares of created values in terms of total economy; its allocation and distribution (based on communal—societal ownership and needs) that is still relevant as a historical inspiration.

The question of social ownership might be the crucial in terms of furture degrowth policies. Although there is a tendency and a strong stream of revisionism by

[22] Such as: production, investments, labor, product distribution, prices, economic laws, management, decision-making, democracy, allocations on amortization, accumulation and consumption, as well as the distribution of personal incomes.

attributing negative narratives about Yugoslav identity through ideology of socialism, which resulted to ambivalent attitudes regarding workers' self-governance by positioning it as an example of dissonant heritage, Yugoslav, it is one of the crucial examples of commoning in practise. On the other side we have romantic streams of filo-Yugoslavs that seek whatever source they can, to defend the idea of unity. I hope that this chapter will be seen as the one that open the critical issues by brinding the crossings between the history and the future to contribute the creation of radically nedded turns on policies, politics and future life-centred development, that seeks for well being, quality of life, equity and equallity. Since the degrowth is a political slogan with theoretical implications (Latouche, 2009), it refers to an "equitable and democratic transition to a smaller economy with less production and consumption" (Martínez-Alier, Pascual, Vivien, & Zaccai, 2010; Kallis, 2011; Simonis, 2010; Jackson, 2011). Degrowth holds layers of theoretical roots (cultural, economic, environmental, political and societal) that overlap and are usually considered trans-disciplinary knowledge integration that inspires future systemic solutions (Kočović De Santo, Milanović Rodić, Babović, Kuzma, 2022). Its cultural roots are related to a radical critique of the development paradigm, where the economy is seen as an autonomous sphere that controls the production of meaning and sense, and it seeks the "decolonisation of the imaginary" that has to be led by the cultural sphere (Latouche, 2014). People are faced with processes of instrumentalization on a personal level by becoming objectified within rationality theory – as the means of a productivist-consumeristic mechanism. The mentioned processes function as a technocratic catastrophe that tends to commodify relations among humans and between humans and Nature (Martínez-Alier et al., 2010; Hausknost, 2017; Illich, 1973, Kočović De Santo, Milanović Rodić, Babović, Kuzma, 2022). Degrowth is also rooted in critical theories, branches and streams such as CS, CDS, decolonisation, and post-colonisation - that share many atributes with self-governance in Yugoslavia.

2.4 The Lessons from Self-Governance in Cultural Policy of Yugoslavia as Endogenous Know-How

Studying historical phases of cultural policy will bring the broader context and highlight the importance of tracing the "place of culture" (Hawkes 2004) and its significance in development. Cultural policy in the history of Yugoslavia had three significant phases (Đukić 2010) that strongly influenced today's perception of the field of culture and the final contemporary sphere of culture.

- 1st phase—before the emergence of Nation-states at the end eighteenth century and 19thcentury, including period after the emergence of Nation-states until the WWII,
- 2nd phase after WWII until the beginning of the 1990s with associated sub-phases

- 3rd phase, from the beginning of the 1990s until the year 2000, transition in blockade" or culture of quasi-nationalism, 4th phase culture in transition or culture in transformation, from the beginning of 2000 until the 2011,
- 5th phase "back to national unity" from 2012 until today.

I gave these phases as tentative, involving Đukić (2010) and Dragičević Šešić (2018) visions and explanations—as most prominent cultural policy authors in Serbia. The special interest is to bring the deeper understanding on 2nd phase, with associated sub-phases, because by involving this historical framework it is possible to bring the deeper understanding on:

1. Self-governance in Yugoslavia as our common dissonant heritage and know-how (specific in terms of market oriented socialism),
2. Institutional terms and conditions,
3. The context of cultural policy knowledge form Yugoslav self-governance—as semi-peripheral and Global South, invisible in Western knowledge that is still relevant and "calls" for the actual decolonization of knowledge.

2.4.1 The Second Yugoslavia Phase

The second phase in Yugoslavia started after IIWW and lasted until 1990. This cultural policy phase is characterised by different (sub-phase) variations of *state-bureaucratic and quasi-state models* within the newly formed after-war Socialist republic. In terms of government, cultural policy moved from centralisation to democratisation. The second phase ended by de-etatisation and decentralisation in culture through three main sub-phases (Đukić 2010), while Dragićević Šešić analyses each sub-phase as independent (phase)[23].

The first (sub) phase—*Agitprop or Social Realism as Repressive Cultural Model*—lasted until 1953. It was characterised by the advantage of revolutionary ideas and values, before actual knowledge about culture and art appeared. During the Agit-prop, "cultural policy was centralised, state governance included many strategies of repression and censorship that led to self-censorship. The communist party formed tasks in the field of culture conducted by the apparatus for *agitation and propaganda*" (Đukić 2010). Social realism appeared as Stalin's model copy of culture in the former USSR. The function of culture, in an ideological sense, was utilitarian and did not encourage the idea of culture as a field for individual freedom of any sort.[24] Luckily, this phase was brief, followed by a period of progressive cultural action (Dragičević Šešić et al. 2018).

[23] Social Realism and a Repressive Cultural Model (1945–1953), Democracy in Culture (1953–1974), Decentralization and Self-Governance (1974–1989).

[24] The actual contribution of this phase was reflected in educational process led by culture, which in the period of its duration included the elementary literacy of 800,000 out of a total of 1,200,000 illiterate people (Dojkinović Đukić 2003).

The second (sub) phase is the *Self-governance* or the *Democracy in Culture (1953–1974)*. The main idea was to achieve democratisation in culture through decentralisation and the introduction of the quasi-state institutions. Two parallel cultural developments can be identified; one was still under strong state and ideological control, and the other was the actual appearance of artistic freedom (Dragičević Šešić et al. 2018).[25]

The third (sub) phase *Liberalisation* or *Decentralization and Self-governance (1974–1989)* followed the previous (sub) phase, institutionally assumed the disintegration and gradual abolition of etatistic centralised structures, with the parallel construction of the new social mechanisms based on broader democratic foundations of consensus and social participation. It is particularly known for the specific policy initiatives to decentralize culture throughout the former Yugoslavia (Dragičević Šešić et al. 2018). Determined also as *Quasi-State Self-Governance* phase where cultural policy was driven through *Self-Governing Interest Communities of Culture* (originally *Samoupravne interesne zajednice* SIZ) as newly formed institutions. It was market quasi-state governance, in which some sectors provided, while others received (financial support/cultural contents). Moreover, both sides were state-imposed through the SIZ system as bureaucratic organisations in charge of allocating resources from the real economy towards cultural organisations. After introducing self-governance communities of interest "*free labor exchanges*" facilitated closer links among cultural institutions and profit organizations in Serbia. The entire cultural, societal and economic system was transformed during this period.

The critical question was whether the culture met the genuine interest in the society as a non-productive sphere from such an insured financial distribution system. The *Cooperative Labour Act (1974)* recognized productive and non-productive organizations of total economy. That reinforced this cultural model of free labour exchange, whereby *non-productive organisations* were also recognised as *labour organisations*. Culture was not separated from other economic activities, but presented instead as the unity of the total economic and societal system, where the cultural contribution was reflected in the identity of workers. Even the legislation recognised the distinction between productive and non-productive labour, with each type recognised as *socially beneficial*.

These two sub-phases were characterised by liberation from etatist influence. Many decisions previously made subjectively were left to the market, where labour organisations become carriers of simple and extended reproduction. The functions held by the state were transferred to *quasi-state bodies*, (in some cases the professional *apparatchiks*) who were representatives of the workers and labour organisations in given segments of the overall economy and society.

---

[25] By the end of the 1960s and beginning of the 1970s, many new institutions and prestigious international festivals for different art forms had been established. A large network of municipal cultural institutions, such as houses of culture, libraries and cinemas was also created. At the same time, many individual artists were sanctioned and their works (films, theatre plays and productions, books, etc.) were banned. This was not an officially proclaimed policy but was exercised through political and ideological pressure (Dragičević Šešić et al. 2018).

It is essential to mention that during these two sub-phases, besides the administrative-budgetary system, three systems of financing culture have been introduced *the market mechanism,*[26] *the system of social funds*[27] *and the system of direct self-government negotiation*[28](Madžar and Popov 1968). Science, culture and physical culture in the period 1966–1972 were financed through the social funds and budget funds (DPZ). With the institutional decentralization after 1974, culture SIZs appeared. Financial resources flow was allocated from the contributions of personal incomes of workers, organizations of associated labour and citizens. The basic difference brought about by the constitutional change in 1974 was reflected through the allocation contributions from associated labour organizations towards the SIZ (the contributions from workers' incomes existed since 1965) (Ristić 1982). With the Constitutional amendments in 1974, "de-budgetization" as economic decentralization, became institutionally supported through the launching of SIZ. Financing the common needs enabled the contributions from the labour organizations. Thus, the necessary financial resources were transferred from the economic sector to the non-economic sector. Earlier sources included social funds, in addition to administrative-budgetary ones. Social funds administered allocation of labour surplus. The social essence of SIZ was reflected through socialization and liberation from work from etatist and market influences. In this way, socio-economic links were established between individual and relatively independent parts of the total economy based on cooperative work, which contributed to the development of an integral system of self-management through agreements, agreements and togetherness.

The previous point regarding politics and ideology is about their reflections on the economic instruments of cultural policy and vice versa. In other words, based on how and what is financed, it is possible to recognise and "read" which value-ideological framework is nested in culture (society and economic policy) even when the explicit formulation is missing as a specific ideology (Kočović De Santo 2021a, b, c, d). The critical, and at the same time radically progressive idea was introduced that the "*development of the culture in the socialist society of workers self-governance in Yugoslavia could be fully realised only through a free market approach*" (Madžar and Popov 1968). It was the market socialism that many scholars such as Bideleux admired by explaining that Yugoslavia took significant steps "towards market socialism in the 1960s" (Bideleux 1986), as an "impressive

[26] The idea was that labour organisations in the field of culture should start earning income, in the same way as working organisations in the economy, depending in part on the culture market trends.

[27] Social funds were pumped directly or indirectly through various instruments (taxes or contributions). They retained part of the budget system, but differed in terms of the manner of allocation. The allocation was essentially different from centralised budget administrative logic. Allocation was conducted by quasi-state bodies competent to trace cultural development.

[28] In the system of direct self-governing bargaining, in which all parties interested in the development of specific areas and branches of culture or specific cultural actions participate. It was conceptually conceived as a deeply democratic process, because in the process of negotiations on the side of the recipients of cultural services—people who allocated money were able to express social needs towards cultural organisations as the providers of cultural services.

headway" (Griffin and Bideleux 1986) by the virtue of the workers self-governance model (Kočović De Santo 2021a, b, c, d).

I found it essential to make further crossings also in this part, by including economic views on the role of the culture at the end of the 1960s in Yugoslavia. Correspondingly, I will intensively refer to extraordinary pioneer research in culture "The Culture as an Activity and Creation in the Conditions of Commodity Production" conducted by macroeconomist researchers[29] in 1968 which starts as "culture is an integral part of social being, and cultural development is one of the most important dimensions of comprehensive social development" (Madžar and Popov 1968).

After the IIWW times in Yugoslavia were characterised by the focus on economic theoretical and practical research. The emphasis was mainly on the political economy, economic policy, and instruments—the material aspects of social reproduction in a narrower sense. Since the 1960s the public finances were studied more intensively, when traditionally unprofitable activities started to appear by being recognised as "the main contributors—in a broader sense—to the overall (social) development" (Madžar and Popov 1968).

The culture slowly came into focus to support and bring the deeper understanding of the very picture of development. The culture in Yugoslavia at the end of the 1960s was understood as an "integral part of the social being" (Madžar and Popov 1968). In mentioned times, the cultural development represented one of the few most critical, but crucial dimensions of social development (Kočović De Santo 2021a, b, c, d). From the economic standpoint, culture was a pivotal factor in the entire developmental process in Yugoslavia workers' self-governance (Kočović De Santo 2021a, b, c, d). Moreover, if "the liberation of people and labour" was the ultimate goal of socialist development, the importance of the "culture as the fundamental infrastructural base" becomes more obvious (Madžar and Popov 1968; Kočović De Santo 2021a, b, c, d).

The cultural researches from 1960s were primarily occupied with the value, ideological and aesthetic aspects.

From early 1970s, scholars in the field of culture and linguistics expanded their scientific interests towards Management in Culture and Arts, Cultural Policy, and other areas in terms of official education syllabuses. Economic aspects of the culture, knowledge of the cultural market, actors and institutional changes, from the end of the 1960s and beginning of the 1970s, would emerge in the coming decades in an interdisciplinary manner, as topics that are not losing their relevance.

As Milena Dragićević Šešić[30]—world's recognized expert and professor in the field of cultural policy notes, the Yugoslav (Belgrade and Zagreb) schools of cultural policy in the sixties and the seventies were in line with the world's leading theories

---

[29] Madžar, Popov and group of authors from the Yugoslav Institute of Economic Research, now—The Institute of Economic Sciences in Belgrade.

[30] Milena Dragićević Šešić was asked as one from 40 professors of cultural policy around the world (The International Journal of Cultural Policy Vol. 16, n. 1) to write review on the book that was essentially important for her—regarding the cultural policy but to exclusively choose from Anglo-American scholars and theoreticians from Northern Europe translated to English. This call has additionally "centralized" true knowledge and produced new peripheries and semi-peripheries of knowledge production.

and practices of cultural policy, *but semi-peripheral status of Yugoslav science, along with the insufficiently committed academic culture of remembrance, as well as new trends of academic colonialism*, completely sidelined the Yugoslav knowledge on cultural policy and self-governance from academic currents (Dragičević Šešić 2018).

Here some facts deserve to be clarified to contribute the decolonization of knowledge. Yugoslav science failed to be included in the Western body of knowledge. For example, *the fact that the organization of cultural activities (cultural management) as an academic discipline was first developed in Yugoslavia in 1961—well ahead of the City University in London, which started in 1967 – could not be found in the texts of Western authors, who still name the City University as first, even though a number of articles had been published in English mentioning the Belgrade school's past and current teaching of the history of art management* (Ibid.). Yugoslav public figures and intellectuals, (mostly Serbian and Croatian), dealt with the most general questions of culture and cultural policy, *from cultural rights to the decentralization of culture*, as a necessary input of the total democratisation processes in Yugoslav self-governance of society, culture and economy. *Stevan Majstorović, initiated founding of the Institute for Studies in Cultural Development in 1967. As soon as in 1968, they launched the journal Kultura (whose first editors were Stevan Majstorović, Slobodan Canić, Dragutin Gostuški, Trivo Inđić, Vujadin Jokić, Danica Mojsin, Mirjana Nikolić, Nebojša Popov, Bogdan Tirnanić, Milan Vojnović and TihomirVučković) to stimulate national authors to research and write but also to translate the most important world theorists such as Leo Lowenthal, Pierre Bourdieu, Frederic Jameson, Ernst Bloch, Herbert Marcuse, Raymonde Moulin, Noam Chomsky, Edward Sapir, Robert Escarpit, Pierre Belleville, Lucien Goldmann, Bronislaw Malinowski, Michel Foucault, Ruth Benedict, Paul Ricker, Charles Snow and many others* (Dragičević Šešić 2018).[31] The cultural policy as academic discipline world-wide was not constituted until the nineties (University of Warwick, 1994), when the first specialized journal was launched—*European Journal of Cultural Policy* (after 1997 it was called *International Journal of Cultural Policy*), or more precisely until 1998, when the first world conference of researchers in the field of cultural policy was organized (*ICCPR - International Conference of Cultural Policy Research*).These details are important

[31] The debates from the new generation of social-humanistic researchers (Stevan Majstorović, Zagorka Golubović, Trivo Inđić, Miloš Nemanjić, Ratko Božović, Branka Prnjat, Sveta Lukić, Radoslav Đokić, Miladin Životić, Sonja Licht, Branimir Stojković, Ružica Rosandić, Bora Džuverović; intellectuals of different disciplines operated in at least three different circles: (1) within the Praxis group (Lešaja and Medak; Lešaja http://www.praxis-arhiva.net/), some of whom were Milan Kangrga, Rudi Supek, Vjekoslav Mikecin, Gajo Petrović; (2) around the journals *Kulturni radnik* and *Naše teme*, and especially connected to the Institute for Development and International Relations (IRMO), like Biserka Cvjetičanin, Nada Švob Đokić, Matko Meštrović; and (3) experts gathered around the projects of the Croatian Institute of Culture—Božidar Gagro, Zdenka Gjanković, Ivan Rogić, Andrija Mutnjaković, Jasenka Kodrnja, Miroslav Tuđman, Ivan Jakopović, Zlatko Sudović et al.) covered a wide range of topics, from the freedom of artistic expression, the development of audiences and tastes linked to the new populist cultural patterns, to the specific cultural policy issues related to the methods of financing culture, art markets, the organisation of cultural life and decentralisation, community centres, cultural centres and festivals of culture, to the relationship between culture and tourism (Ibid.).

to compare with the fact that there was the journal *Kultura* in Belgrade in 1968, as '*the journal of the theory and the sociology of culture and cultural policy* (Ibid.).

It is obvious that Yugoslavia was ahead of the curve in this area, but that it remained globally unrecognized due the work of *epistemic erasures* still present in the knowledge, institutional designs at the local, national, regional, European and international levels—treated as semi-periphery and less important knowledge.[32] Epistemic erasures could be understood as Fricker's epistemic injustices, who define two types: testimonial and hermeneutic injustice.

Testimonial injustice critically questions (to decolonize injustice), the academic imperialistic idea that the lower credibility is being assigned to specific knowledge and sources, due to the prejudice against them, including scholars in the periphery (and other minority groups) (Fricker 2007). This created invisible, but existing academia class system, the semi-periphery and Global South knowledge is treated as "*lower*" to contribute the total knowledge consensus, policy or decision-making, where the argument regarding the deficit of credibility is rooted in the prejudice by representing fair explanation on the Dragičević Šešić (2018) insights on invisible and forgotten Yugoslav cultural policy, and we can say that the same applies to self-governance as a Yugoslav endogenous knowledge in terms of theory and practice.

Hermeneutic injustice means that marginalized communities are unable to give meaning and intelligibly articulate their experiences to dominant groups. The essence of this is explained by the work of incommensurability as such communication is based on different value and narrative systems, where community's experiences often fall into a conceptual blank gap and cannot be explained or translated through the concepts and approaches that are produced by the dominant groups (Fricker 2007). These concepts are valuable framework to understand the power, and the processes how the different linguistic and ideology apparatus, reveals different interests, by leading to incompatible systems of measurements (e)valuation mechanisms.

The main issue regarding self-government in the field of culture was the "concentration of bureaucratised power that arose from the status of SIZ as the only mediator and arbiter in the labour exchange process, explained as the so-called double state" (Đukić 2022). Đukić (2022) explained that despite theoretical differences between cultural policy models, in practice and strategic decision making processes in cultural development there was, and still exist" a symbiosis of bureaucracy in the public administration and networks of political parties, civilians organization, media and intellectual, cultural and artistic—"*Elite*" that operates within a "*dissident sub cultural model*". Thus, in the field of culture, she called it a "double state", because it restrains and restricts, instead of encouraging cultural development, why it cannot

[32] The only translated book—*Cultural Policy of Yugoslavia*—was published in French and English, as a part of the first globally significant project to map cultural policies, implemented by UNESCO in preparation for the World Conference on Cultural Policies in Mexico in1982. Neither Stevan Majstorović nor most of the researchers who closely dealt with the field of cultural policy had been published in international journals before, since the main form of international cooperation was through state projects of bilateral exchange and multilateral cooperation (UNESCO) (Dragičević Šešić 2018).

be called true cultural, but more pseudo-cultural, which is just a step away from anti-cultural policy" (Đukić 2022). Some authors challenged (sub)phases two and three as (none) *democratic times*,[33] bearing in mind the one-party system, the developmental directions from the 1960s until 1980s in Yugoslavia entered the space of decentralisation, which assumed institutional separation from the state centralised concept towards participation in governance within workers self-governance. Polšek explained the last decade of self-governance in culture as a state of systemic artificial demarcation line between the economy and the non-economy as "a fake, ideological synthesis", where the process of social consensus, or the social law in socialism is understood through the elasticity of class demarcation line, "where the role of science, education in general, culture, mass media, etc. in socialism, was inversely proportional to economic prosperity", and the culture was politically liberalised, but economically suppressed (Polšek 1989).

In the field of economic theory and cultural policy, many researchers still seek to reaffirm the idea of Yugoslav self-government "corrected and adjusted to the new situation and context", " as a contribution to the history of degrowth and its possible future engagement" (Kočović De Santo 2021a, b, c, d; Nina Mihaljinac 2021). However, this model of cultural policy established in the early 1970s by all characteristics presented an innovative attitude of the state towards the management of not just cultural, but total social development in more integrative manner.

2.4.2 The Third Yugoslavia Phase

The third phase in Yugoslavia started after 1992 and lasted until the year 2000. The cultural policy model was a variation of the *state-bureaucratic* in a socialist republic, where politics and policies were centralised again. This cultural policy model is called *Agitprop of Third Yugoslavia,* where the third Yugoslavia after the civil war narrowed to the territory of Serbia and Montenegro. The positive contributions achieved from the past were abolished. The basic idea was to agitate the "national" values given by the political elites for the newly created state of Yugoslavia (Serbia and Montenegro), which represented discontinuity from the previous period, similar to the period after WWII. The fall of the Berlin Wall was an important historical event that announced world changes. The disappearance of the *Iron Curtain* assumed global political and ideological transition of a formerly large number of socialist countries towards neoliberal capitalism. While the world witnessed globalised mass transitions, Yugoslavia suffered the civil war. These circumstances delayed transition until the 2000s. According to Dragićević Šešić, the 1990s are the period characterised by the centralisation of culture. The cultural policy did not contain a defined concept or priorities, starting from attitude towards institutions at the municipal or republic level to the lack of agreements on programmes and financial plans of cultural institutions, incompetent managing bodies (Dragićević Šešić 2002). This grim situation lasted

[33] The major argument for challenging democracy was the one-party system.

until political change happened at the local level in 1997, and intensified after 2000. These reforms helped the painful process of transition, which started after 2000th when the private and civil sectors took their more substantial participation role in each aspect of social life after two decades.

The current cultural policy is determined by the historical context of belated transition, transition in progress, and post-transition challenges reflected in each segment of society and the system. The modern post-transition context in Serbia defined it as a country in *permanent crisis*[34]that shares attributes with other "more developed" ones, in terms of objective systemic uncertainties determined by neo-liberal institutional frameworks (Kočović De Santo 2021a, b, c, d). The period from 2000th to date, has not brought support for the possibilities of creating genuinely democratic institutions, which would support cultural values and concepts to contribute political actions to become (again) more socially constructed. On the contrary, in Serbia, modern dynamics political-party changes have been frequent since the 2000s, and they not necessarily reflecting ideological differences. In the observed period, political changes have brought shortened discontinuities determined by political mandates. Seemingly, neoliberal tendencies, with the existence of political conservatism sets a new demarcation line between the national values (identity, culture, tradition, religion) and European values, with significant control over the media with national frequencies and a program that brought passivity in society while separates it from the real-life and critical thinking (Kočović De Santo 2021a, b, c, d). Media transition brought the "realitification manner" in everyday life, through intensive imported reality-show concepts, spreading values far from virtue. Also new media identities are rising in the digital environment, whose role and relations with traditional identities of the place shall be deeply investigated. The process is mainly about reducing *politics to the polite interplay between diverse cultural groups* (Kapoor 2008).

On the one hand, it is an ongoing stimulation through the definition of current policies towards preserving national cultural identity, which becomes suffocated through the overemphasised neoliberal market discourse and narratives in the public sphere, to a certain extent answering the needs of the power holders. The atomic bomb seems worse, but its adverse effects last for shorter time compared to the vacuum of permanent crisis with provided eternal transition from one system to another, which is the universal issue of most of post-socialist transition countries.

It is possible to comprehend the strong relationship between political economy, ideology, system, state and culture even without further analysis. The connection between the governance options is strongly related to the political economy of any current condition. Almost every cultural policy is carried by different interests that could be perceived as expressions of the ruling class's narrow ideological, political and economic objectives at the state level, and beyond by the work of international bodies' factual involvements and contributions. Based on author Ferguson's *anti-politics machine,* focused on elements of power discourse that led to development

[34] That holds attributes of: country in a race to the bottom, developing country, country of the Global South and periphery.

interventions with unexpected adverse outcomes, explaining how "independent and objective" evaluations from World Bank launched development recommendations by strengthening the influence of political elites of the state through delegated implementation of requirements (Ferguson 1996) as the "ever-changing forms of neo-colonial hegemony" (Kapoor 2008). Foucault's thoughts are targeting the power-knowledge relations; are useful to have on mind in inspiring decolonised knowledge "what is needed is a study of power in its external visage, at the point where it is in direct and immediate relationship with that which we can provisionally call its object, its target, its field of application, where it installs itself and produces its real effects; In other words, individuals are the vehicles of power, not its points of application; The individual which power has constituted is at the same time its vehicle" (Foucault and Gordon 1980).

In future inspirations on new forms of participatory and direct democracy decision-making, the potential shortcomings that have arisen in historical context and practice should be considered to mitigate more profound destruction instead of humanising modern systems fraught with problems.

Still, it is possible to conclude that the progressive and transformative potential of the culture is excellent as its role towards change. Previously is supported by Bourdieu's view on the evolution of capital into the *symbolic capital*, where he pointed out the essential role of the *language, symbols and social properties* of each part of the total economy (Bourdieu 1972).

2.5 Discussion: For the Future Built up on the Decolonized Knowledge Integration—Towards the Life-Centred Development

The preceding section brought us to the second point of the chapter, where I intend to bring the trans-disciplinary integration of knowledge by involving deeper analysis on natural aspects of culture. This part refers to eco-feminist, heterodox and evolutionary economic approaches and sustainable market transitions, which represent a logical continuation of the previous - focused on the historical analysis of cultural policy in the context of the critical evolution of development. Building theoretical and practical direction enables modelling the new systemic concept for the future fairer life-centred development.

How to achieve practical alternative, starting from critical narratives far beyond modernisation theory, towards fair development? Is this about degrowth in practice (Kočović De Santo 2021a, b, c, d) as the other side of the decolonisation of development and knowledge creation (Smith 2012; Kočović De Santo 2021a, b, c, d)?

Post-colonialist lenses brought perspectives on cultural politics, seeing it as a dynamic process of meaning-making that engenders the construction and contestation of power relationships (Kapoor 2008). The formal decolonisation happened and was followed by the creation of postcolonial nation-states, decolonisation of minds, in terms of "escaping colonial mentality" (Kočović De Santo 2021a, b, c, d) including "long-term process involving the bureaucratic, cultural, linguistic and psychological divesting of colonial power" (Smith 2012), as an ongoing process until today.

People, their interrelations, value creations and expressions, in the context of living nature, lead us to the necessity of reimagining the development by putting life in the centre of planning. This process is about the decolonisation of the contemporary meaning of sustainable development to mitigate all recognised obstacles. Hudson discusses the decolonising act as a process that is "*about to engage with meaning-making practices critically, as these underpin the most fundamental aspects of being human; it, therefore, helps remember Collini's goal of extending human understanding through words, ideas, narratives the art, artefacts that help us make sense of our lives and the world we live in*" (Hudson 2020).

The model of the new systemic solution shall bring the decolonised understanding of "*economics as nothing more than public policy in the function (and as a reflection) of society, environment and culture*" (Kočović De Santo 2020a, b).

Involving *care* for the people, community, and environment is the vital contribution from the feminist political ecology in building up alternatives based on strata of different societal needs. Harcourt summarised post-capitalist ways to "*live with and redefine capitalism aware of social and ecological limits*" to change our economic values by including "*care*[35] *and respect for our families, communities, other knowledge and cultures*" (Bauhardt and Harcourt 2019). Harcourt proposal is about redesigning economies "so that life is valued more than money" (Bauhardt and Harcourt 2019). Eco-feminist approaches given from Rao and Ahram-Lodhi intended to explain the relation between capitalism and culture where "*capitalism is not just as an economic system, but an institutional order that shapes the culture, polity, as well as the economy*", by prioritising the accumulation of surplus instead of focusing on "*life-making*" (Berik and Kongar 2021).

Moreover, life-centred development implies greater participation of people in culture and politics, achieving genuine democratisation through the socialisation of institutions. Aiming to address what is the good life, Dengler and Seebacher, through feminist-degrowth view, explained that "*degrowth add resses all of us as active, society-shaping citizens, rather than consumers*", by putting as its core "*autonomy and participatory, collective bottom-up decision-making processes*", where the essence of "*quality of life might lie in needs to be collectively re-evaluated with an emphasis on 'conviviality'*" (Dengler and Seebacher 2019). Conviviality, understood as sociability, is defined by Ivan Ilich (Ilich, 1971). He aimed to point to reflection outside the framework of industrial production, toxicity, and violence at extractive limits where the people become commodified by technocratic chaos (Kočović De Santo 2020a, b). At the same time, Mbembe recognised conviviality as a point of reflection on people and their participation in deconstructing colonial African

[35] Is about how communities organise their community and their livelihoods; Care is about looking after and providing for the needs of human and nonhuman others; it is about the provision of what is necessary for good life; is about unpaid work that needs to be done as externalised care, etc.

countries within authoritarian regimes (Mbembe 2001). More recently published, *The Second Convivialist Manifesto* is against neoliberalism, productivism, and populism; It values relations of cooperation that allow humans to compete without hubris and violence by taking care of one another nature[36] (International Convivialist 2020).

Future life-centred development is necessary to increase the understanding of care as equally crucial for the contemporary economy as it reflects conviviality (Kocovic De Santo 2021a, b, c, d). Care is usually unpaid work. Dengler and Strunk offered the conceptual framework towards an alternative that might help overcome the boundary between "*the monetised and the maintaining*" economy by promoting "*a revaluation of paid and unpaid care work*" (Dengler and Strunk 2018).

Life-centred development can be explained through critical eco-feminist vistas as unifying human-nature duality "*through the interaction of culture, history, discourse, technology, biology, and the 'environment,' without privileging any one of these elements*" (Gaard 2018).

How to achieve life-centred development, how to finance it, and how to maintain good life? Kočović De Santo explained as follows: *(1) Decolonising knowledge on development as degrowth in practice, (2) Progressive political transformation, 2a) Macro and micro institutional evolutions* (Kocovic De Santo 2021a, b, c, d) the life-centred development involving mentioned reference is shown in Appendix C.

Involving heterodox economics approaches is crucial as it gives alternative economic insights except the neo-classical (Petri 2021; Wray 2020; Arestis and Sawyer 2019). The focus is on Mallor's work since she offers participatory options towards democratisation of the money as a publicly shaped phenomenon and good to support fair future sustainable transition in the wake of the most significant identified challenges (Bauhardt and Harcourt 2019). She stands for public money by explaining that it creates the framework for private profit[37] (Mellor 2015). "Money is a social phenomenon and a public resource", which "with the collusion of the state, has been harnessed for the benefit of a privileged few". Mellor devoted her work to explain how money is created and where it comes from, to deconstruct the neoliberal narratives that "money is made[38] in the commercial sector "and taxed into the public sector.[39] By pointing that public expenditure often comes before the loans

[36] Recognises three main effects of conviviality as responses of social movements and groups towards the system, (1) in more developed countries social movements of young people respond to environmental issues; (2) in the decades of authoritarian regimes, social groups of young people are organised towards the systemic change; (3) in transition countries from post-authoritarian systems to developed countries, social groups of young people–migrate.

[37] If environmental sustainability and social justice are to be achieved, public money must be reclaimed to support social and public provision of goods and services in the same way it has been used to save the financial system.

[38] Two sources of money creation widely recognised: bank lending and public spending, and other forms of newly created local social currencies and crypto-currencies. Effectively mentioned 2 sources led to the banks' privatisation of the creation of the national currency through the creation of bank account loans.

[39] Under neoliberal ideology, the state must not create its own money; it must rely on taxing the wealth-creating sector, that is, the market. Mellor pointed to the fallacy in this argument: by the fact that the public sector also pays taxes.

and taxation. Moreover, she offered the basic income for care work. "*If new money emerges through debt in the commercial sector, extracting tax income will always be uncertain; If a basic income were to be funded by state-issued money, its implementation would depend on the expenditure priorities of the state*" (Bauhardt and Harcourt 2019). Mellor opened crucial questions regarding the sufficiency, social justice, and providing enough goods and services to maintain a good quality of life that the money system needs to ensure, where the economic priorities would be determined in the interests of the most vulnerable members of the community. This is a suitable way to re-think the financing of other non-profitable, but spheres full of sense, like culture and environment, that goes in line with her comparison "*while commercial views of money will always stress value in money terms as profit*" (money invested in making more money), "social and public forms of money can address outcomes in terms of social and public benefit", where the market value and judgement becomes replaced with social value, judged in terms of personal relationships (Bauhardt and Harcourt 2019).

By merging technological and innovation opportunities in the light of the Fourth technological revolution understood as a "*link between global climate change and evolving socio-technical landscapes*"(Rip and Kemp 1998) with historic lessons regard the positive aspects of workers' self-government in Yugoslavia[40]; The culture (with "allies from activism and third sector"(Köhler et al. 2019) becomes the "carrier of sustainable transition markets" and the sphere of common need, started from the niche markets, providing the radical change in terms of development, by the "*functioning of alternative, informal, community-based markets that arise by collective contributions*" (Boon et al. 2020) led by the cultural force as control factor of the systemic imbalances, within the decentralised and direct democratic system.

Speaking of dichotomies developed versus developing/underdeveloped; centre vs periphery; Global North vs Global South, the conclusion is that historically less privileged countries in hegemonic disadvantage are dealing with permanent pressure. The question is: Can they change their position in any way?

In terms of capturing transition routes towards life-centred development, identification of present context and historical lessons gave us material for the interpretation of possible future directions as a more desirable way that actively involves culture. Building a conceptual proposal in light of sustainable transition markets and technological change allows applying essential insights regarding the radical changes and innovations in development. Radical innovations appear in the specific context[41] that enables (previously made) knowledge to be launched (Rip and Kemp 1998). They may endanger current activities and become rejected and unwanted since they replace or at least fundamentally modify existing paradigms (Rip and Kemp 1998). Furthermore, relying on Rip and Kemp, "if governments or societies desire a new technology" or paradigm, "they must construct its artefacts and create a transition path toward it" (Rip and Kemp 1998). The sustainability transitions literature "recognised

[40] As inherited endogenous knowledge for decentralisation, democratisation and participation in politics and policies.

[41] Wartime, crisis etc.

the importance of civil society and social movements" in transforming our production and consumption systems by building support for transition policies and providing protective spaces for innovation that have "less obvious effects on broader cultural values and beliefs" (Köhler et al. 2019). The importance of grassroots practices and bottom-up approaches recognised as "the main allies in societal change" who are bringing attention "to justice, fairness" by creating "semiotic maps of the possible and desirable shifts" led by values is obvious, emphasising that social movements, especially when they are engaged with industrial change, can bring attention to the needed cultural change (Köhler et al. 2019).

Jessop pointed on the dependence of the activities of social agents and the broader conjuncture "even if capitalist bias is deeply embedded in the matrix of a capitalist type of state, the latter's policies could work against capital as inimical forces capture the state apparatus or pressure it to pursue capitalistically irrational policies" (Jessop 2016). Relying on the previous, equilibrium is about balanced economic strategies, state projects, and hegemonic visions to support policies and politics for the society. Therefore, if actual democratic values are adhered to, life-centred development shaped by culture and nature must hold the central attention.

Kate Raworth gave an exceptional framework of Economic Doughnut (Raworth, 2017) for future development that can be explained as a radically innovative macro and micro-organisational and analytic framework, from the view of sustainable market transition theory. "Radical innovations are assumed to emerge in niches, where new entrants (pioneers, entrepreneurs) nurture the development of alternatives" (Rip and Kemp 1998). The radical innovations may breakthrough more widely if landscape developments put pressure on the regime that leads to cracks, tensions, and windows of opportunity (Köhler et al. 2019), explaining how the interactions between niches and regimes occur on the multiple dimensions and confrontations while navigating transitions.[42] The Economic Doughnut is an excellent point to go further on development, towards the life-centred development. The group from IPE[43] went further by involving other vital indicators in Degrowth doughnut as a reshaped version to reflect the construct of a safe, just, and sustainable degrowth transition.

Cultural development implies the continuous progress of human activities and lives in culture, art, and other areas. The needed re-conceptualisation of the life-centred development process as integrative towards a gender, indigenous people, authentic culture, race, justice and equality, equity in terms of culture, societal environmental, and economic, energetic issues is the strategic task to deal with in the future. These challenges have been recognised as suitably integrated within degrowth (with permanent feedback from social movements practices), a centrally important starting point for future research.

---

[42] *Markets, regulations, cultural meanings, technologies and are enacted by interpretive actors that fight, negotiate,*
search, learn, and build coalitions.

[43] Institute for Political Ecology, Zagreb, Croatia http://ipe.hr/en/degrowth-donut/.

To build an alternative model of fairer development for the future, I proposed decolonisation of development, focusing on endogenous knowledge of workers self-governance in Yugoslavia as commonly inherited, which offered the last historically noteworthy democratisation decentralisation in decision-making politics. Speaking in terms of proposals for the future, the first challenge to overcome is to achieve democratisation in policy and politics. On the one hand, in the contemporary context, it is about indirect mechanisms of decision making, which is exercised by political power holders. On the other, the direct participation of people in the governance of their own and political life assumes the existence of institutions and high conscious regard for the development processes in society.

As such, the self-governance in the second phase of Yugoslavia represented an institutional, inherited and endogenous knowledge as a historical lesson that supports the future life-centred development by seeking substantial democratisation of political processes and practices of participatory governance models—that could be upgraded for the current context. I referred to theories out of the mainstream knowledge zone, following decolonising knowledge by escaping "proven given best practices" of development.

Inspiring future fairer development is about putting the life into the heart of development *that* comes as a reflection from the previous parts—in light of bringing true sense and meaning into the policies. In short, as a synthesis, it is about spreading the solidarity from human–human and human-centred (including human rights, and beyond) towards human-nature to overcome anthropocentrism. This direction will allow the broader discussion on the meanings of the good life and on the valuation of life more clearly as the phenomenon shaped by joint work of culture and nature. Cultural and natural aspects define life. By the way, we live our everyday life, express ourselves, and articulate (distribute) values through social networks, within the natural context—of permanent life creation, values and meanings of the ecosystem. The commons and values discussed earlier in the text should be the core, fundamental point to project everything else - from economic, developmental, policy and politics important for the life, each should be done as a participatory process. Finally, the idea of modelling future fairer development that is life-centred recognises the progressive and transformative role of culture towards participative governance, decentralisation, and democratisation of institutional and political processes to achieve and maintain the good life. Our common *habitués* culture and environment are contextualised by place consisted by for reflections of living and silent voices to be integrated through life—in heart of the future development.

When we think of natural aspects of culture and cultural aspects of nature, life is the actual denominator of both, as a bridging point of sense. As shown in Fig. 2.2, future policies and politics should start from the two-heart(s)-reflections—as life-centred crossing that integrates natural and cultural context. See also Appendix C as the synthesis of the previous research.

The decolonisation towards development as fairer progress for the vest of majority is about the life-centred practice. It shall be the quest for a laborious process of cultural and knowledge decolonisation. It is a process of intellectual, ethical, emotional and spiritual decolonisation, which can offer the essence for life outside the zones of

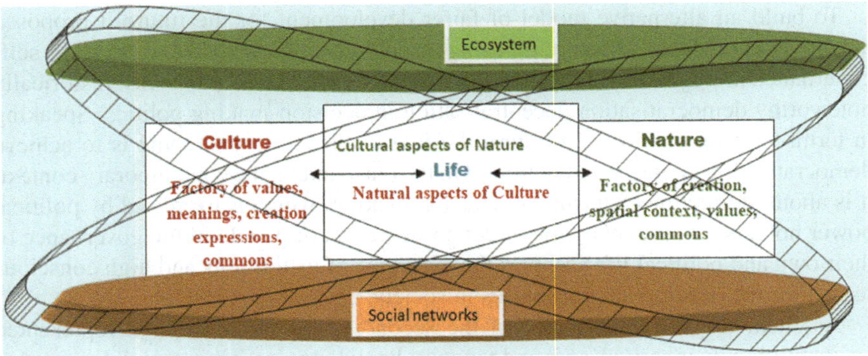

Fig. 2.2 Conceptualising future life-centred development by inspiring life into the heart of the processes, Author contribution

consumerism and productivism. It unequivocally paves the way to overcoming capitalism. The changing paradigm can reflect the main idea on a personal, societal and state level of behaviour in terms of involving the following states as axiomatic to crash the undesirable matrix on many levels.

(1) The only thing that has no alternative is life—replace the universalism towards "pluriversalism", to achieve possible alternative worlds;
(2) Reduce—to live, and let others live—become the activist consumer. An activist consumer is a person who is responsible for consumption. Buy if needed. By choosing products we buy, we directly invigorate national economies, choose large corporations or small producers, and greater or lesser environmental footprint involved in the production. It is the way to express the sensitivity towards the hegemony of each type of exploitation, but also the way to support the economies of the Global South through careful product selection;
(3) Protect the local production—if it is underdeveloped, let it mature! On the state level, there are instruments to protect the market by tariffs and substitution that works in the short-term by enabling the domestic production to mature. Internal production and consumption lead to industrial development and expansion, where organising system into networks of cooperatives gives opportunity to maintain achieved industrial growth. Also there are available instruments to protect the labour and work within the development countries (universal basic income, universal basic services, job guarantee, working time reduction etc.). Similar scenarios were tested and recognised, at least in the short run, as desirable in developing and underdeveloped countries.

2.6 Conclusion

The analysis of institutional changes in culture and economy in and from the 1960s onward is an exceptional basis for interpretation(s) of policies, ideology, and market and the inspiration for further modelling fairer development. The aim of chapter was to (1) bring the critical reflections and understandings on endogenous forces and possibilities versus exogenous influences as essential for the future life-centred development (led by the culture that involves the system of values and meanings towards the reconstruction of ideology from the sphere of culture).

Regarding the first point of the paper, based on the theoretical literature overviews, historical views on self-governance in Yugoslavia and the phases on cultural policy (CP), culture travelled through different stages of cultural policy models. Different sub-phases and CP models describe the second Yugoslavia. They were starting from state centralised and planed CP, within *state-bureaucratic*, towards *state-prestige and quasi-state model* - when the shift from the centralised state towards cultural unification within socialist Yugoslavia happened.

The democratisation of culture and later cultural democracy, through decentralisation, appeared within the *quasi-state model* of cultural policy within self-governance in socialist Yugoslavia. Second Yugoslavia witnessed delegation of powers from the state bodies to expert bodies. We noticed relations between economic and non-economic—the sphere *common* needs, where the cultural system developed. It was characterised by specific allocation policy and direct self-governing bargaining, in which all parties interested in the development of specific areas and branches of culture participated. People could express their social needs towards cultural organisations, as *liberalisation in culture.*

The third phase of Yugoslavia was characterised by the reversible process of strong centralisation and *state-bureaucratic model* to re-build the newly Yugoslav national identity (Serbia and Montenegro)—given from political elites towards populous.

From the year 2000 when political, ideological and systemic changes happened, we are dealing with the transition CP model. The current CP model reflects cultural confusion and environmental crisis, which shares attributes and challenges with a modern economic system—of permanent crisis faced by countries in hegemonic disadvantage due to the mentioned global dichotomies. Speaking of contradictions developed versus developing/underdeveloped; centre versus periphery; Global North versus Global South, the conclusion is that historically less privileged countries in hegemonic disadvantage are dealing with permanent pressures and crisis, seeks for another developmental solution. These issues are the primary source of future uncertainties.

By crossing historical analysis with reflections from the cultural political economy, critical cultural and developmental studies, in terms of institutional context, practice, shared values and total governance system—regarding the *place of culture* helped in conclusion that the culture as one sphere of common needs and common consumption in the 1970s was on the pedestal in Yugoslavia, by enjoying its significance as a combination of logical consistencies derived from ideology, public practice, politics, and its recognized contribution to the social reproduction and the total development. This helped us understand the strong relationships among political economy, ideology, system, state and the place of culture. The links among the governance options are strongly related to the political economy of any current condition. Every (cultural) policy is carried by different interests that could be perceived as expressions of the ruling class's narrow ideological, political and economic objectives at the state level and beyond—by the factual involvements and contributions of international bodies that contributed to the state power. We could conclude how mentioned historical lessons and actions affected a change of values and shift away from profit-seeking as a single production objective toward a concern for non-profit in terms of total development. That is important for contemporary issues and needs of integrative approaches towards cultural and natural development that respects their uniqueness and diversity.

This research quest seeks and confirms that profound democratisation of our societies is essential by giving the meaning to human life outside of consumption and materialism, through the struggle for the good life, a concern for equality between living beings and their needs (Kočović De Santo 2021a, b, c, d; Latouche 2004; Kothari et al. 2018).

In these terms, by decolonising knowledge and development, I made deeper analysis towards arguments out of mainstream and comfort knowledge zones. This approach enabled the trans-disciplinary integration of knowledge (CDT, CCS, Degrowth, eco-feminist political ecology, heterodox and evolutionary economics, sustainable transition markets) that recognised endogenous knowledge, values, meanings, care, pure senses—of life—are made by the common work of culture and nature.

The shown historic findings and lessons from the second Yugoslavia and workers self-governance as an endogenous knowledge that represents the starting point for future systemic changes, inspired mostly by many scholars degrowth. Involving endogenous knowledge as cultural, it was possible to conclude that the progressive and transformative potential of the culture is supreme as its role towards change, having in mind the power of culture for future fairer developments. In this culture launches the development—as a reflection of social and not political elites construct.

As I mentioned, we need to put life at the heart of development. I was reflecting on Hawkes and beyond, where the role of *great cultural change* is exceptionally seen as progressive and transformative. Moving on from historical lessons of role of the culture in the total economy, system, and vice versa, involving more explicitly findings gave the argumentation for conceptualising life-centred development. Summarising my previous research presentation shown in Appendix C, future life-centred development shall start as an overlapping process of (1) decolonisation of knowledge as practical work of degrowth and culture as an expression of values, meaning, and sense; (2) political progressive change towards direct democracy, autonomy, participation and socialisation of institutions in work for the society and environment, where inherited endogenous knowledge is very welcomed by offering a deeper understanding of the specific context in the selected country of hegemonic disadvantage, (3) followed, and (at some point) pushed by institutional evolution (economic doughnut, platform cooperatives, in light of sustainable transition market idea).

The conducted historical analysis helped us identify the role of the culture in future life-centred development, which enlightened the context depended on "sub-periphery and Global south knowledge". Therefore, if actual democratic values are adhered to, life-centred development shaped by culture and nature must hold further the central attention.

Acknowledgements This chapter partially involves my previous research findings as a fundamental basis for this research contribution. It is related to the ongoing projects:

COST Action 19,129 Decolonizing development: research, teaching, practice;

Decolonizing Degrowth—beyond the Eurocentric Western nature of conceptualizations and movements, supported by University of Łódź, Poland;

EPICA—Empowering Participation in Culture and Architecture: Activating Public Resources for and with Local Community (ID 7,744,648), supported by Sicientific Fund of Serbia.

This chapter reflects the conclusions from my and other authors' research presented in 2021. where we all met within two International Degrowth conferences—*Building Alternative Livelihoods in Times of Ecological and Political Crisis International Online Joint Conference of the international degrowth research networks, the International Society for Ecological Economics, and the European Society for Ecological Economics, University of Manchester, UK, July 2021* and *Caring communities for radical change, Hague, August 2021.*

Also, my chapter relies on some additional previous research findings presented at the *International Conference on Conviviality*, 2021 New Zealand, and the *National Conference on Institutional changes in the economy of Serbia through history*, organized by the Centre for economic history within the Institute of Economic Sciences, Belgrade, Serbia.

Finally, I express my special gratitude as editor of this special issue to co-editor Stephanie Eileen Domptail, all the authors and many colleagues (reviewers) that enriched this publication with their priceless suggestions and work. As an author of this chapter, I express my special gratitude to Professor Inga Kuzma as the first reviewer of my chapter, and Professor emeritus Milena Dragićević Šešić UNESCO expert and great connoisseur of world cultural policy, who contributed to this chapter and the publication as a whole with review work and excellent suggestions.

Appendix 1

Life-centred development

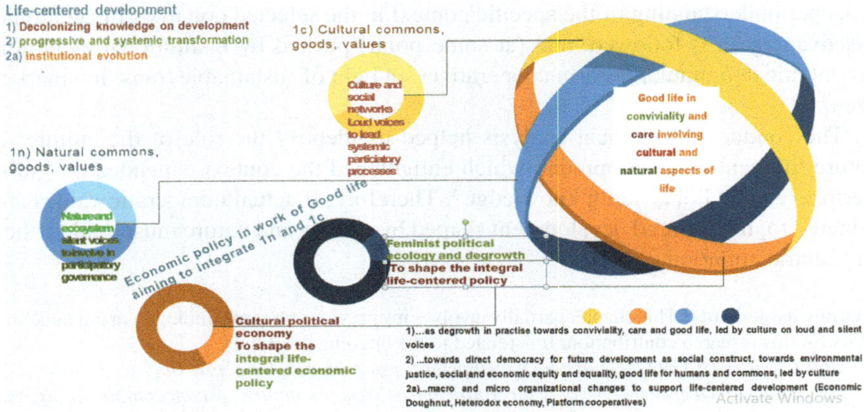

Source Kočović De Santo Presented at International Degrowth Conferences organized in July University of Manchester (2021) *Eco-cultural Aspects of Degrowth: The Story of Decolonisation of Colonial Mentality (2021a)* and Hague (2021) *Escaping Development Hegemony Towards the Life-Centred Development* (2021b), and International conference on conviviality (2021) *Social movements as living labs of conviviality and their (in)formal transnational context in Balkan* (2021c)

References

Albritton R, Jessop R, Westra R (eds) (2007) Political economy and global capitalism: the 21st century, present and future. Anthem Press, London, New York

Althusser L (1970) Ideology and ideological state Apparatuses. In: La Pensée (ed) Lenin and philosophy and other essays. Mon Rev Press 1971. http://www.csun.edu/~snk1966/Lous%20Althusser%20Ideology%20and%20Ideological%20State%20Apparatuses.pdf

Arestis A, Sawyer M (eds) (2019) Frontiers of heterodox macroeconomics. Springer International Publishing, Cham. https://doi.org/10.1007/978-3-030-23929-9

Bauhardt C, Harcourt W (2019) Feminist political ecology and the economics of care: in search of economic alternatives, 1st edn. Series: Routledge studies in ecological economics. Routledge, New York. https://doi.org/10.4324/9781315648743

Berik G, Kongar E (eds) (2021) The routledge handbook of feminist economics. Routledge, Abingdon, Oxon; New York, NY

Bideleux R (1986) Ch 11: Yugoslavia, Hungary and the Vicissitudes of market socialism. In: Communism and development, 1st edn. Routledge, London, p 328. https://doi.org/10.4324/9781315780627/communism-development-routledge-revivals-robert-bideleux?refId=a07a9c36-85d5-4536-a6e0-11b039c0d657

Boon WPC, Edler J, Robinson DKR (2020) Market formation inthe context of transitions: a comment on the transitions agenda. Environ Innov Soc Transit 34:346–347. https://doi.org/10.1016/j.eist.2019.11.006

Bourdieu P (1972) Outline of a theory of practice. Cambridge University Press

Castro CJ (2004) Sustainable development: mainstream and critical perspectives. Organ Environ 17(2):195–225. https://doi.org/10.1177/1086026604264910

Christoforou A, Laine M (2014) Re-thinking economics: exploring the work of Pierre Bourdieu. Routledge, Taylor and Francis Group. https://www.routledge.com/Re-Thinking-Economics-Exploring-the-Work-of-Pierre-Bourdieu/Christoforou-Laine/p/book/9781138226760
Dengler C, Strunk B (2018) The monetized economy versus care and the environment: degrowth perspectives on reconciling an antagonism. Fem Econ 24(3):160–183. https://doi.org/10.1080/13545701.2017.1383620
Dengler C, Seebacher LM (2019) What about the global south? Towards a feminist decolonial degrowth approach. Ecol Econ 157:246–252. https://doi.org/10.1016/j.ecolecon.2018.11.019
Dragićević Šešić M (2018) Culture and democracy: cultural policies at work; Is cultural policy theory possible at the semi-periphery (Kultura i demokratija—kulturne politike na delu—da li je moguća teorija kulturne politike na poluperiferiji?), J Kultura, (no. 160):13–35. Link to English version, https://www.casopiskultura.rs/wp-content/uploads/2021/09/prevodi/6.%20Milena%20Dragicevic%20Sesic,%20CULTURE%20AND%20DEMOCRACY,%20Culture%20No.%20160,%202018.pdf
Dragićević Šešić M (2002) The creation of the myth of the city, the policy of spectacle. Public Cult, Policy Magna Agenda, Beograd
Dragićević Šešić M, Dragojević S (2005) Management in turbulent environment. Clio, Beograd
Dojkinović Đukić V (2003) Transitional cultural policies confusions and dilemmas. Zadužbina Andrejević, Belgrade
Dragićević Šešić M, Tomka G, Mikić H (2018) Compendium cultural policies & trends, country profile Serbia, 19th edn. https://www.culturalpolicies.net/wp-content/uploads/pdf_full/serbia/serbia_092018.pdf
Đukić V (2010) Država I kultura: studijesavremenekulturnepolitike. Institutzapozorište, film, radio I televiziju-Fakultetdramskihumetnosti
Đukić V (2022) Duplicate of the State in the cultural policy of Serbia (1971–2021), Serbian politica thought 1/2022. https://doi.org/10.22182/spm.7512022.5
Escobar A (1992) Imagining a post-development era? Critical thought, development and social movements. Soc Text (31/32):20. https://doi.org/10.2307/466217
Ferguson J (1996) The anti-politics machine "development," depoliticisation, and bureaucratic power in Lesotho, 3rd edn. The University of Minnesota Press
Foucault M, Gordon C (1980) Power/knowledge: selected interviews and other writings, 1972–1977, 1st, American. Pantheon Books, New York
Fricker M (2007) Epistemic injustice: power and the ethics of knowing. Oxford University Press.
Gaard GC (2017) Critical ecofeminism. Lexington Books, Lanham
Griffin K, Bideleux R (1986) Communism and development. Econ J 96(381):240. https://doi.org/10.2307/2233448
Hawkes J (2004) The fourth pillar of sustainability: culture's essential role in public planning. Cultural Development Network: Common Ground, Melbourne
Hawkes J (2013) Shaping policies: culture-sensitive and context-based policies in sustainable development, presented at the international congress, "culture: key to sustainable development", Hangzhou, China. http://www.unesco.org/new/fileadmin/MULTIMEDIA/HQ/CLT/images/Jon_Hawkes_social_cohesion2.pdf
Horvat B (1969) An essay on Yugoslav society. Intcrnational arts and Sciences Press Ine White Plains, New York
Horvat B et al (1970) Economic functions of Federation. The White Book, Report. Institute for Economic Sciences, Belgrade
Hudson H (2020) Larger than life? Decolonising human security studies through feminist posthumanism, Strateg Rev South Afr 40(1). https://doi.org/10.35293/srsa.v40i1.269
International Convivialist (2020) The second convivialist manifesto: towards a post-neoliberal world. Civic sociology, June. https://doi.org/10.1525/001c.12721
Jessop B, Sum NL (2006) Beyond the regulation approach. Edward Elgar Publishing. https://doi.org/10.4337/9781845428907
Jessop B (2016) The state: past, present, future. Polity Press, Malden, MA

Kapoor I (2008) The postcolonial politics of development. Routledge, London; New York
Kocovic De Santo M (2020a) Growth and degrowth: economic pillar as a threat to the cultural and ecological sustainability of public and common goods (Commons). Book Proc Inst Radio TV Theater Film Fac Dram Arts Belgrade, vol 36
Kočović De Santo M (2020b) Chapter: shifting of economic cycles conditioned by non-economic factors—hope for solutions from the periphery. In: Black Swan in the world's economy. Institute of Economic Sciences, Belgrade. https://www.library.ien.bg.ac.rs/index.php/monog/article/view/1296
Kočović De Santo M (2021a) Ch 2: Historical lessons on institutional changes in the department of culture. In: The importance of institutional changes in the economy of Serbia through history edited by Jelena MinovićMilicaKočović de Santo Aleksandar MatKović. Institute of Economic Sciences, Belgrade, Belgrade
Kočović De Santo M (2021b) Eco-cultural aspects of degrowth: the story of decolonisation of colonial mentality. Paper presented at the building alternative livelihoods in times of ecological and political crisis international online joint conference of the international degrowth research networks, the international society for ecological economics, and the european society for ecological economics. University of Manchester, UK. https://www.academia.edu/45594532/Eco_cultural_aspects_of_Degrowth_The_story_of_the_decolonisation_of_colonial_mentality
Kočović De Santo M (2021c) Escaping development hegemony towards the life-centred development. Hague. https://cloud.degrowth.net/s/gbP6C3wz9BH8Aj9
Kočović De Santo M (2021d) Social movements as living labs of conviviality and their (in) formal transnational context in Balkan. https://perc.ac.nz/wordpress/wp-content/uploads/2021/10/Conviviality-Program.pdf. https://perc.ac.nz/wordpress/ideologies-tools-and-advocacy/
Kočović De Santo M, Milovanović Rodić D, Babović D, & Kuzma I (2022). Empowering strategic priority areas of cultural policy: empirical findings on cultural participation processes in Serbia. Strategic management: Int J Strateg Manag Decis Support Syst Strateg Manag. https://doi.org/10.5937/StraMan2200030K
Köhler J et al (2019) An agenda for sustainability transitions research: state of the art and future directions. Environ Innov Soc Transit 31:1–32. https://doi.org/10.1016/j.eist.2019.01.004
Kothari A, Salleh A, Escobar A, Demaria F, Acosta A (2018) Pluriverse: a post-development dictionary. https://degrowth.org/2018/04/14/new-book-pluriverse-a-post-development-dictionary/. https://www.academia.edu/39692614/_Pluriverse_A_Post_Development_Dictionary_AUF_2019_NEW_BOOK_edited_by_Ashish_Kothari_Ariel_Salleh_Arturo_Escobar_Federico_Demaria_and_Alberto_Acosta_Download_full_ebook_for_free_PDF_License_Creative_Commons
Latouche S (2009) Farewell to growth. Polity, Cambridge; Malden, MA
Madžar LJ, Popov S (1968) Culture as an activity and creation inthe conditions of commodity production. Yugoslav Institute for Economic Researches (Institute of Economic Sciences)
Marcus G, Fischer M (1996) Anthropology as cultural critique an experimental moment in the human sciences. https://press.uchicago.edu/ucp/books/book/chicago/A/bo3613216.html
Mbembe A (2015) Decolonising knowledge and the question of the archive, p 29
Mbembe A (2001) On the postcolony. University of California Press, Berkeley
Mellor M (2015) Debt or democracy: public money for sustainability and social justice. Pluto Press. https://doi.org/10.2307/j.ctt18gzdmr
Mihaljinac N (2021) What do we mean when we say. In: New cultural policies. Institute for Philosophy and Social Theory (IFDT), Belgrade
Munck R (2016) Critical development theory: results and prospects. https://www.researchgate.net/publication/315729676_Critical_Development_Theory_results_and_prospects
Pieterse JN (1996) The development of development theory: towards critical globalism. Rev Int Polit Econ 3(4):541–564. https://www.jstor.org/stable/4177204
Petri F (2021) Microeconomics for the critical mind: mainstream and heterodox analyses. Springer International Publishing, Cham. https://doi.org/10.1007/978-3-030-62070-7.

Polšek D (1989) Cultural policy in SFRY. https://www.academia.edu/6219764/Kulturna_politika_u_Jugoslaviji_UNESCO_1989_
Raworth K (2017) Doughnut economics—seven ways to think like 21st century economist
Rip, A., & Kemp, R. (1998). Technological change. In Rayner, S & Malone, E.(Eds), Human choice and climate change: 2, Resources and Technology, (pp. 327-399). Battelle Press.. https://kemp.unu-merit.nl/Rip%20and%20Kemp.pdf
Rist G (2010) The history of development: from western origins to global faith, 3rd edn. Zed Books, London
Ristić, Ž., (1982) Self-managing system underlying the financing of the public expenditure. In: Branislav Šoškić (ed) Problems of the self-managing economic system, upon the 30th anniversary of self-management in Yugoslavia, vol II, pp 559–593
Sachs W (1992) The development dictionary, p 353
Smith LT (2012) Decolonizing methodologies: research and indigenous peoples, 2nd edn. Zed Books, London
Šoškić B (1982) Chapter 1: specific historical features and problems of the self-managing economic system. In: Šoškić B (ed) Problems of self-management economy system, Montenegrian academy of science and arts, vol 1, upon the 30th anniversary of self-management in Yugoslavia
Sum NL, Jessop B (2013) Towards a cultural political economy. Edward Elgar Publishing. https://doi.org/10.4337/9780857930712
Wray R (2020) A great leap forward, 1st edn. Elsevier, Cambridge

Dr. Milica Kočović De Santo works as a senior researcher at the Institute of Economic Sciences in Belgrade, Serbia. She holds a graduate diploma from the Faculty of Economics in Belgrade (2009). She defended her Ph.D. thesis (in January 2017) within the department of Cultural Policy Management in Culture and Arts at the Faculty of Dramatic Arts. Her interdisciplinary research with trans-disciplinary conclusions is positioned in the theoretical triangle of economic cultural, degrowth theories, and decolonization studies. She teaches at the Faculty of Dramatic Arts and University of Arts, Belgrade, Serbia at MA and Ph.D. study levels.

Chapter 3
Anticipation of the Degrowth Concept in the Socialist Republic of Poland of the 1970s

The Case of the Artistic and Scientific Movement

Inga Barbara Kuźma

Abstract The paper discusses the concepts of degrowth and decolonisation in the context of the tensions of the Cold War in socialist Poland of the 70-ies. At that time, the Berlin Wall symbolised the divisions between two geopolitical structures, namely the West, perceived as a guarantor of democracy, and the East, dominated by the Soviet sphere of influence. However, both sides were full of anticipation, and tried to predict the directions the world would further develop facing the era of industrialization and consumption, which were perceived as serious environmental threats. In the West, these concerns were voiced by the degrowth movement, and empowered by the decolonization processes. In the East, such criticism was expressed by some artists and intellectuals individually interpreting and presenting the 'western' news in an eastern environment. Noteworthily, the East was more actively following the intellectual trends emerging in the West than vice versa. Hence, the eastern and western awareness of critical reactions to global changes was neither spread evenly nor symmetrically. The paper aims to illustrate the creation of discourse addressing systemic criticism of development using Poland as an example of emerging locally interpreted postulates of changes, referred to as civilization, progress, growth, development as well as the changes in human behaviour patterns. The discussed case is presented from the angle of decolonisation of knowledge adopting a post-dependence approach as the author's perspective and suggesting a more reflective term for this case, namely Decolonization Degrowth Turn.

Keywords Decolonization Degrowth Turn · Decolonization of knowledge · Critical post-development approach · Post-dependence theory

---

I. B. Kuźma (✉)
Institute of Ethnology and Anthropology of Culture, University of Łódź, Łódź, Poland
e-mail: inga.kuzma@uni.lodz.pl

Łódź, Poland

M. K. De Santo and S. E. Domptail (eds.), *Degrowth Decolonization and Development*,
https://doi.org/10.1007/978-3-031-25945-6_3

3.1 Introduction to the Authorship Approach

The paper is inspired by the decolonisation of knowledge approach, which originated from the disagreement on perceiving western knowledge systems as being hegemonic. This trend triggered the need to construct other knowledge systems by researching alternative epistemologies, ontologies, and methodologies (see Broadbent 2017; Hira 2017; Wood 2018).

Importantly, this approach is not only typical of the so-called Global South, but as the author believes, also of the geopolitical construct referred to as the East, or Second World versus First World meaning the West (Solarz 2009; Sztompka 2005). The current nomenclature puts it in a post-dependence region including Poland. Due to her education and professional experience, the author herself supports this standpoint (theory) (see Harding 2004).

However, reaching this point and education has been an ongoing process for the author of this paper since its late start. The reason for this was that the author's ethnological and anthropological study did not include post-dependence theories. They began to emerge in Polish Humanities around 2000, when she was already completing her MA graduation in 2001; the author's Ph Thesis 2001–2006 did not follow a top-down program model, but one especially created to cater to individual needs. The individual choice of the program allowed the research of such concepts as dependency, domination, or hegemony as well as discourse status in a political context, where knowledge meant power (see Laska 2011; Howarth 2008). The most invaluable source of inspiration for the author's study was the literature by Michel Foucalt.

Interestingly, the first step towards a post-dependence approach observed in Polish Humanities was translating the post-colonial theories to the experience of post-socialist countries, as illustrated in Thompson's work (Thompson 2000), one of the first of this type studied in Polish academia and translated into Polish. However, according to Anton Saifullayeu, there was resistance against recognising Middle and Eastern Europe as post-colonial regions. This was supported by Western academic circles who did not accept Russia and Germany as the colonisers of Europe (see Saifullayeu 2015: 78).

However, according to Saifullayeu, several researchers, including the above-mentioned Ewa Thompson and other Polish scholars claimed that excluding Russia from being a coloniser, was the consequence of what was accepted by western European intellectual circles. Specifically, they accepted such Russian immigrants" interpretation (following the October Revolution and then WW2) of Russia as being a traumatised and suffering country (also victorious in WW2) rather than aggressive and dominating.

Another reason noted by Saifullayeu was the lack of colonial discourse in Russian historiography as well as the scarce attempts to create post-colonial discourse in the historiography of the nations invaded by the Russian Empire, and as he claims further, 'The lack of discourse means the lack of problems' (Saifullayeu 2015: 78; see Hairong 2007: 444). Also, Guayatri Spivak, a post-colonial theorist claims that post-colonial analysis of Middle and Eastern Europe, must follow different research patterns, as each region needs individually established research criteria and Dariusz Skórczewski, another Polish researcher cited by Saifullayeu, notices that 'one of the main reasons why post-colonialism entered the academic research and debates in Middle Europe relatively late, is that this post-colonial theory is philosophically and methodologically entangled in Marxism, which was a disgrace to the societies [in Central and Eastern Europe—I.B.K.] and therefore hard to accept by the Polish researchers as the subject of study [...] due to the totalitarian experience'] (Saifullayeu 2015:78).

Polish post-dependence can be referred to as 'an umbrella term encompassing a group of significant and institutionalised articulative practices addressing human experience, identity projects, social, political and cultural relations, axiological and symbolic community imaginary worlds, types of perceptions following, and at the same time influenced by the period of dependence' (Nycz 2011: 8). According to Nycz, the opposite of post-dependence is not an absolute form of independence, but various forms of dependency and configurations of dependency relations. This creates opportunities for the emergence of control, subordination, and incapacitation mechanisms, and at the same time, the acts of choices, auto-creation or inven-tive actions beyond the existing and dominating structures (see Nycz 2011: 8).

The author of this paper believes that in such a situation individuals fall victim to subordination and dependency at the level of attitudes or mentality, while their experiences and trauma may fade in the past. The individuals entangled in such situations clearly will attempt to take actions aiming to extricate themselves, thinking and acting against the imposed norms by subversion. What needs to be emphasized here is that taking a counter approach by the individuals originating from post-dependence society can undergo different forms of transformation of such experiences. On the one hand, the foundation for this was laid by the revolt against the CCCR, the biggest hegemon of the Soviet bloc, where the communist regime spread its interpretations of freedom, democracy, and citizenship. This, along with the disagreement and revolt, triggered the admiration for the West, which partially could have been seen as the symbol of independence and truth, countering the Russian regime. On the other hand, and which does not contradict the former, the subversion was, in the time of the regime, to be based on a creative take on western values. Such inspirations resulted, for example, from the lack of understanding of their original context and attempts to enter dialogues or debates from different, individual perspectives, facing specific experiences of totalitarian dependency.

Grassroots subversions in the first dependent, then post-dependent regions, could, in the author's opinion, follow another path. Specifically, their development could have been based on the experience of hardly tangible news, on either side of the Cold War, and they became a perception from a distance.

Such multiplicity, as the author believes, encourages auto-reflection still operating within post-dependent groups. It can be invaluable to explain the dependent past of one's own group, applying noticing and appreciating own subversive past actions and the attitude towards specific present values.

The reflective approach adopted to interpret past events benefits from introspection, thus helping to realise that differentiating between progress and regress, success and failure can be never straightforward. In the author's eyes, in the process of "progressing" democratisation or "liberating" the economy from being capitalist or neoliberal, any progress-related issues included in discussions entered by post-dependency cultures are typical of these regions. Such cultures must struggle to self-position themselves by counterarguing from their inferior positions, while their functioning in the past was based on being dependent on and in the opposition to the other side, while their positioning was built on constant comparisons.

3.1.1 Is the Post-Dependent Decolonizing of Knowledge Possible?

Differentiating between the East and West can never be zero–one. The eastern perspective, as the author believes, reveals a high level of self-relativisation that does not exclude showing superiority towards the dominator. The sense of being superior despite objective sub ordinance, whether a formal, legal, or political one, addresses not only the authority, but also the culture from which the authority originates. This explains the hundred-year tensions between Poland and Russia, or Poland and Germany.

Therefore, the Polish language developed a variety of terms, carrying negative connotations and idioms aiming to ridicule the Russians and their former Soviet rule, including the Polish supporters of and those actively participating in PZPR (The Polish United Workers' Party), referred to as Katsaps, Russkis or party officials. One of the famous jokes circulating at the times of the Polish People's Republic was for example, 'Let the German occupy you and the Russians expel you' expressed to curse somebody or wish bad. Another example can be 'Whom shall we hit first, if Poland was invaded by the Germans and Russians, as in 1939? Germany—Why? Because first duty, then pleasure' (Głos Wielkopolski 2019).

At first sight, these jokes were told in a relatively light-hearted spirit. However, it needs to be noted, that the penal code of 1969, fully abolished in 1998 clearly stated that imprisonment from 6 months up to 8 years was the punishment for the insult, ridicule, and humiliation of the PRL (The Polish People's Republic) or any state organs in power. The authorities had the freedom to interpret these offenses very broadly. For example, the author's grandfather was deprived of his prosperous job and moved to a less prosperous position for telling jokes about Stalin. Yet, he was treated sympathetically.

However, making constant references and comparisons results in 'subtle reification of divisions [...] even with splinters of cultural—neo-colonialism by uncritical imports from Western theories (Nowicka-Franczak 2017: 404). Therefore, it can be assumed that decolonization of knowledge embeds critical thinking within the West and auto-reflection on one's own contextualising in the East and the effects of being under the influence of the First World.

This kind of post-dependency thinking, practised by individuals culturally and biographically linked with the former eastern side, can bring results for a valid reason. It allows, as the author believes, to distance yourself from one's own western suburban inferiority complex, as the West was seen in the East as the 'true' source of wisdom, and broadly understood progress, innovation or ample opportunities. Such an approach meant self-subordination. Further, the author claims that adopting such an approach—choosing to remain the worse can act against oneself, who really can never follow and always remain behind as the Other. One can feel such alienation in the context of the East that in its own eyes becomes the other, i.e. the worse, just by not belonging to the West.

As noted by Michał Buchowski a Polish Anthropologist of culture and an ethnologist: Said, Wolff and Todorov's concept of orientalism and perception of the process of creating the other were consistent with a modernistic vision of the world where the Berlin Wall was an emblematic symbol [...]. In this politically shaped and conceptualised world, orientalism connotes the East that can be distant, central, close, or even European. Regardless, it always denotes a place homing its inhabitants and culture, all being essentialised' (Buchowski 2006: 46).

According to Buchowski, the transformations taking place in 1989, in the European context triggered dismay, chaos, uncertainty, cognitive dissonance, symbolic disorder, a stoppage in a ritual, which had to be experienced. Following the old paradigm, Buchowski claims that the most feasible solution was moving the Western borders further to the East. Hence, further orientalising of further cultural regions took place.

The author of this paper advises the researchers to consider the opportunities created by post-dependency studies while using decolonising of knowledge postulates in their research. This encourages one to auto-critical reflection and insight into one's own internal conditioning. In the case of post-dependent cultures, such conditions are the response to the influential trend, originating both from the regime and the cultural and political order, and considered as superior. Such a response can be not only manifested by revolts, leaving dependency, or adopting new order, but also by seeking new possibilities.

This is consistent with the assumption of the concept of degrowth that is not limited to an alternative economic approach. As noted by Gorgos Kallis, Federico Demaria and Giacomo D'Alisa, this critically approached concept was invented to practise decolonising the public debate of 'the idiom of economism' and cease to treat 'economic growth' as a social objective (see D'Alisa et al. 2020: 36–58).

The degrowth concept embraces the idea of decolonisation (meaning changes in the realms of imagination following Latouche 2020: 117), with the main points focusing on criticism of GDP growth, capitalism and evoking capitalist ideas of social systems, as well as commodification.

The criticism of these areas does not fully exploit the potential of degrowth, as it is also a source of positive notions, advocating welfare, recovering, and creating the common weal, transformation towards reformed work and production organisations and based on values sharing (see D'Alisa et al. 2020: 40).

The main point of the criticism from the above-mentioned authors is the concept of 'development' connotated with culture-rooted meanings, understood as a way of behaving to progress towards a predetermined and arbitrary destination.

The author believes, consistently with Serge Latouche, that 'decolonising imaginations' is essential to successfully address one's own ideological, political, economic and social entanglements. They also shape the content of knowledge and the ways it is transferred and implemented. If they result from personal choices or public debates, they are hard to criticise. What should be noted here is, that the process of decolonizing of knowledge should involve discussing the content of such knowledge, its roots and conclusions as well as its results and outcomes. In addition, one should not underestimate and uncritically look at the starting point of the perspectives adopted while creating and distributing knowledge, but look critically at it and inspect its 'hidden curriculum'. Decolonization of knowledge the more increases in value the more its auto reflective and auto-critical potential is stressed, which has also a political tinge.

3.2 Decolonization Degrowth Turn

The author of this paper emphasizes the importance of the approach proposed by Kirsten Hastrup, an anthropologist of culture, namely, the turn (Hastrup 2008). The turn, understood as 'turn towards/to' means, in Hastrup's take, that discourse draws on first the problem it addresses and shapes this problem at the same time. Second, the researcher introspectively applies one's own cultural principles and perspectives in their research.

Encountering other knowledge, intellectually and emotionally, gaining new experiences at the same time, and referring to previous experiences take place, as a cogniser is far from being a tabula rasa. In Hastrup's turn, there is an inherent connection between cognition of others and self-cognition. To specify further, such an anthropological perspective, establishing any cognitive contact with a problem or a group will involve relativizing. This happens because any cognising subject will relate to one's own principles and even question them to allow analysis and evaluation of other perspectives and solutions, as well as embracing other ontologisms and cultural epistemologies that the cogniser encounters. The author of this paper notes that the understanding or interpreting of Hatsrup's ideas needs to adopt a specific approach. Following Hatsrup's analysis, clearly any attempts to fully understand other social specificity or multiplicity, impose a "becoming" the other, going beyond one's own imagination in order to apply different categories in search of new interpretations.

Only if one can go beyond one's own perception and schemata, can one realise new conditionings allowing necessary space for realising new values, classifications and adjustments. If such restructuring of imagination takes place, even a well-known social reality emerges in a different light and allows different explanations. Ways of thinking embed alternative projects resulting from making specific choices as well as interpretive and auto-critical work. Such choices become informed, dictated by ethics or politics, rather than a researcher's routine, instinct or mere tradition.

Therefore, the author is interested in practising, also attempted in this paper, applying auto-critical and auto-reflective decolonisation of knowledge in the context of post-dependence studies, using a case of artivist activities in Poland in the 1970s. The case has been chosen as the example of demonstrating multiple meanings: not only the subversion against the East, and anticipation of the fascination with the West, but also its interpretation of global trends, undergoing at the same time the process of auto-colonisation of eastern intellectualists, in its niche and original take.

However, the author sees it as a milestone in constructing alternative approaches, important both to decolonisation degrowth approaches, and invaluable in post-dependence studies. The artistic and historical context of the discussed case from the Polish People's Republic, requires consideration of the importance of such terms as development, and associated with it, growth and progress. This also means, as stressed earlier, applying specific human thinking and experience (auto) turn proposed by Hastrup. Thus, it is justified to name this approach Decolonization Degrowth Turn within the post-dependent history.

3.3 General Frame of the Chosen Case

The starting point is established in the 1970s in the 'Global Peripheries', where the Polish socialist system was founded. The core topic of these reflections addresses the Polish "Decolonisation Turn" towards degrowth from the mentioned decade, choosing eco-artivism (more about "artivism" see: Golańska and Kronenberg 2022) as a central phenomenon from its beginnings in Poland.

The 1970s was a critical decade for developing the degrowth concept in the West (D'Alisa et al. 2020). In the West, it was also the time of different minorities' struggles (similar to decolonization movements), an eco-art blossoming (see e.g., Kastner and Wallis 2010), and an increase in the second Wave of feminism (see the popular vision of the western history of this period (The 1970s, 2021). In addition, in 1972 the UN (in the form of the United Nations Conference on the Environment in Stockholm) appealed against the degradation of nature due to human civilization development, as well as social inequalities like colonialism, racism, and exclusion (see United Nations Conference on the Human Environment, 1972).

This era in the Socialist Republic of Poland marked a sudden rise in economic advancement, fueling aspirations for well-being and Western-style prosperity. Hence, Poland was open to Western markets and their models of economic improvement (Winkler 2019; Kościk and Głowiński 2007). From the other perspective, this was the time when conceptualism in art was developing. According to Art and Documentation (2012), the researchers investigating this phenomenon or any definitions of art or artists of the 1970s used the criterion of being a conceptualist or not.

The above-mentioned issue of the journal describes conceptualism as 'a critical tool having avant-garde features, novelty, progress and competition' (Art and Documentation, 2012: 3). Such intellectual storms/ferments resulted in art events originating in the mid-1960s, reinventing the questions about the shape of human and world conditions, and seeking a new formula integrating artists and art critics, or other scholars. The place of such activities was important too, being mainly a landscape or workplace of groups, where nature dominated. The environmental threats posed to nature in the 1970s (along with the increase in ecological issues arising in a global discourse), and the natural environment began to be treated by western and partially eastern societies as strong determinants of individuals' behavior and their creative potential.

3.3.1 "Glocal" Chosen Context

The post–dependency discourse can be identified as the specific "glocal" variant ('glocal' in the sense of a portmanteau of globalization and localism; see more Khondker 2004) of postcolonial studies (Kola 2018: 16). The post–dependency approach is strictly linked to 'border' regions in terms of symbols and systems of values. These regions are referred to as 'Non-Western Europe', 'New Europe', 'two-speed Europe', and, as in the series initiated by Philip Roth, the 'Other Europe'—something in between the original and 'it is a better or worse copy, but only a copy' (Kołodziejczyk 2014: 10).

The former divisions into East–West, The First, Second, and Third World, and others mentioned earlier, were joined by the opposition of Global South and Global North (Cf. Pal 2005; Lees 2020). This has changed the ways of addressing global differences since the end of the Cold War. Such classification could pose a challenge to the self-identification of post-dependent regions in Eastern Europe. In Martin Müller's interesting interpretation, classifying the world into Global North and South did not leave room for what he referred to as the Global East comprising societies like, for example, post-socialist countries and societies occupying interstitial positions between the North and South. He further claims this region to be in a twilight zone, as not clearly belonging to either of the global regions and being referred to as 'the blind spot' (Müller 2020).

Apart from a strong economic division, as in other geopolitical divisions, the demarcation line between the Global North and South is determined by various factors. In common understanding these factors comprise the role of the market and economy, based on existing growth, consumption, large production volumes, and advanced means of exploitation; also, innovation is another important aspect, measured by the freedom index, democratisation, and other values typical of western models of civil society. These values determine the stability of the government systems based on the rules complying with human rights.

These concepts became the starting point for post-dependent regions to implement any geopolitical and economic transformations involving the required legal and public institutions' functioning adjustments. It also involved proclaiming the freedom of speech, which meant the abolition of censorship, and opening media markets and education to a multiplicity of subjects and voices. It also meant opening up to self-organising and non-governmental grassroots movements. These transformations, which took the form of legal policies in Poland after 1989, enabled Poland to enter the EU and other European institutions. Implementing any social reforms was justified by and dictated by Western values, as the West best-personified liberation from a totalitarian regime. However, implementing these changes, especially aiming at equality and non-discrimination, for example, freedom of speech, minorities and gender rights, and respecting women's rights, to name a few, met widespread political and public resistance in Poland of 1989, which was also voiced by 'Brussel Colonisation Discourse' (see: Jabłońska 2009; Budzyńska-Daca and Rusin-Dybalska 2022).

Another factor considered in differentiating between the North and South is their shared past, but differently experienced and remembered depending on the place. The history and dependencies are mainly remembered from the perspectives of the ones who took advantage than from the perspective of the others who were used. This memory is rooted in the author's retrospection of Łódź in the 1990s, where the author lived and personally observed the changes. Łódź was then a city dominated by the textile industry. This industrial monoculture had been shaping the city since the nineteenth century and its collapse in the 1990s resulted in so-called structural unemployment, reaching 9.6% and increasing to 15.4% in 1991, and 20% in 1992. More than 80% of textile factory workers were comprised of women. The lack of employment brought impoverishment to society and the marginalisation of ethical norms. The number of individuals depending on social welfare increased from 28 to 104 thousand in 1989–1992 (Money.pl, 2019).

From the author's and grassroots knowledge perspectives, these tragic real-life stories of suddenly impoverished individuals, including women, were rarely addressed and reported in the world dominated by capitalist and liberal propaganda of success in that time of economic transformation in Poland, instead prioritising individual success. These transformations of the 1990s strangled artificially raised and sustained principles of workers' solidarity and also independent and self-governing trade unions grassroots movements, such as Solidarity from the epoch preceding the political and system changes in Poland.

In the ex-East, the transformation process was contextualised not strictly by economy and politics, but also by identity and culture. The changes occurring in Poland after 1989 opened the minds of society to trends and ideas coming from the West (Wallas 2010: 363). On the other hand, they triggered an identity crisis. The author would like to quote Ilona Copik, summarising Polish sociologists' observations, 'transformation is a global social phenomenon, where the political fact of change in the system just initiates wider processes and their phases. Initially, it seemed to have been a transition from a totalitarian regime to freedom based on the principles of liberal democracy, then from an irrational economy based on the command and distribution model, to the one welcoming the opportunities of the free market' (Copik 2019: 23).

Copik further reminds us of the wave of euphoria over morality, following the fall of the hated regime and the end of "the evil empire". It was sustained by the faith that western success can follow in Poland. However, the implementation of western models and re-joining Europe does not function as a universal norm and capitalist Poland is largely wishful thinking. Poland was a country of contrasts, where on the one hand there were visions of unlimited opportunities, and on the other hand, a very high inflation rate, proclaimed by new media and strength of the individual power, and at the same time, the powerlessness of individuals against unemployment spreading over more and more sectors (Copik 2019).

The price of such transformation was high, as demonstrated in previously mentioned scenes set in Łódź. Consequently, the groups that failed to win by new rules created stereotypical judgments about themselves and others. These judgments drew on persistent sentiments about historical facts, which in the case of Poland, were: 1.5 centuries of the partition of Poland, the trauma of damage suffered from German Occupation during and after World War Two, and further partial Sovietization of social life after 1945, depriving society of full sovereignty (Gosk 2012: 51).

As stressed by Irena Gosk: 'moreover, any reaction of the subordinate to subordination means entering the relationship with the subordinate and leaving marks' (Gosk 2012: 59–60). This researcher highlighted that there emerges a problem of avoiding certain topics and keeping them silent, as breaking that silence and gaining previously unattainable critical insights might pose risks. Specifically, it might reveal the truth that the former world of Soviet domination feeding the fascination with the West shaped the others, formerly dependent on specific east–west domination, who in consequence, partially formed the new type of ex-dependent Polish identity.

Before the development of the discourse in response to the transformations of the 1990s, the language of the 1970s was voicing the decade of apparent development and changes, aiming to stimulate the so-called 'little stabilization' lasting till the late 1960s and following 'The Second Republic' of communist Poland in the 1970s. The former authorities responsible for the 'little stabilization' were replaced by those who were more open-minded to the West, supporting new trends, and introducing 'The Second Republic' as the program and idee fixe adopted by the government in the 1970s. Their promises alluded to creating, in a generation's time, a "better Poland", modeled on western standards, prioritizing the construction of residential areas, improving social welfare, and investing in consumerism. However, these were mere propaganda catch-phrases resulting in awakening social aspirations of a socialist society already suffering from inevitable frustrations triggered by a decade of revolting against the lack of fulfilling fundamental needs (Kościk and Głowiński 2007; Winkler 2019). From the point of view suggested by the author, the government's new policy was a sign of entering a pro-western fascination phase. Due to the credits from the West, the Polish authorities of this epoch, experienced a certain sense of the economical 'autocolonisation'.

Nevertheless, the plans for the development of the socialist economy following the 'Western spirit' were not realized by the government's economic mistakes and the global energy crisis at the same time led to an economic crisis after a decade of decline (Sasanka 2011).

The above situations justify the diversification of discourse applied to communicate the ideas related to specific socialist 'westernization' of the economy and lifestyle (including more liberty) in pursuing dreams of progress and a 'better' life. The aforementioned diversification represented a feature of political and public manifestations, many of which were broken up by armed authorities, with victims on the protesting side. Finally, artistic and intellectual subversive practices significantly contributed to the diversity of the discourse. Like the earlier mentioned open-air event, they often touched on questions that were less obvious than those defined by communist politics, locality-oriented and marking their presence in the discourse.

3.4 The Case of the Open Plain Air in Opolno-Zdrój

The author is interested in art as the scope for the interpretations defined as Decolonization Degrowth Turn in the context of post-dependency studies. The author further believes that blending art and activism into artivism is still one of the most promising tools to implement, and at the same time allows change and its critical evaluation. It is possible due to, as frequently mentioned by the author, the power of auto-reflection, empowered by artivism.

As Dorota Golańska and Anna K. Kronenberg noted, artivism, combining art and activism, stimulates undertaking political and social interventions (Golańska and Kronenberg 2022). Consistent with the noted researchers, the author believes that the practices of the creators, activists and artivists can give a better insight into other knowledge-generating practices, such as think tank, media or artistic ones. These practices addressed processes of knowledge production, and at the same time interpretive practices put forward a specific perspective of seeing the world, taking advantage of the considerable potential of symbolic power they had at their disposal.

What is worth mentioning here, is Osieki, a place hosting plain-air events from 1963–1981 (the last event took place when martial law was imposed in Poland), and the other is one of the earliest interdisciplinary initiatives was the 1st Symposium of Artists and Scientists 'Art in the Changing World' held in Puławy S.A. Nitrogen Plant, in 1966. The initiator and main organizer of the event was a conceptual art theoretician Jerzy Ludwiński, who engaged later in another plain air event in Opolno-Zdrój and other following events. The structure and title of the event were related and relevant to the ideas developing in the 1960s, contextualized by art, technology, science and industry. Importantly, these terms mostly served propaganda purposes, at the same time voicing a strong fascination with technology, cybernetics, and their influence on everyday life. Typically, when industrial plants are the location for such events, they grant access to workshop rooms, industrial equipment and the assistance of the workers. The symposium was then, a platform for creating excellent opportunities for critics and scientists to exchange ideas in discussions and plenaries: see The Archive of Eustachy Kossakowski 1966, also Leśniewska-Zagrodzka 2006; Out in the

Fresh Air? (http://www.fundacjaarton.pl/?ID=1039); Treliński and Pierzgalski 2018; Kalicki et al. 2008; Worłowska et al. 2019; Krukowska and Rzerzycha-Myśliwy 2021. In this trend from Poland of the late 1960s and 1970s, the author proposes to in-scribe another experimental eco-artist and the same scientific open-air event entitled 'Science and Art in the Process of Protecting the Natural Environment from Man', which took place next to the town Opolno-Zdrój in Lower Silesia in 1970s. It is the case chosen by the author herself, to interpret through its lens the question of decolonization of degrowth in the post-dependent context.

This event—in the author's eyes—more than others, represents the anticipation of Degrowth in the East and it is also the realization of one of the first Eastern European (and Polish) regular intersectional critical reflections on the world's future condition, integrating elements of science and humanities. This reflection anticipated future conditions of the human environment, shaped by new and different aspects.

Consequently, that historical Polish eco-activist-scientific artistic actions in Opolno-Zdrój were included in the interdisciplinary symposium as well dedicated to the ecological, artistic, architectural, and research perspectives aimed to address the threats posed to nature by transformation, industrialization and urbanization. The event program was based on assumptions underlining the context of overproduction, consumption, pollution, and humanity's broken relationship with nature (Gołkowska and Ludwiński 1972). Such platforms like this open-air event, were invaluable at that time, as they created rare opportunities to undermine the norms existing in the Polish narrative. Therefore, the language of the discourse applied various means to convey diverse messages.

The open-air of Opolno-Zdrój clearly extended the so-called locality entering the glocal level addressing the following:

1. Intellectual construction and glocality;
2. the pre-degrowth concept was implemented in the 'Second World', which is seen in the 1970s, especially in terms of ecology, but also economic and ethical contexts (criticising overconsumption);
3. questioning the post-dependency hypothesis, the open-air event in Opolno-Zdrój was going in the direction of the independent creating and recreating of degrowth (making its debut on a global scale, including the West, despite political dependence on the USSR, censorship and the economy imposed by Real Socialism).

Among the participants of this open-air meeting there were not only Polish artists, scientists and experts but also representatives of the local community—the inhabitants of Opolno-Zdrój, miners and the management of a nearby mine. This event can be considered a "milestone" at that time, raising the awareness of emerging environmental issues. The event drew attention to the relevant economic issues and posed questions about the 'progress of civilization' (the original world from the documentation of the open-air event; see Gołkowska and Ludwiński 1972) and its role in knowledge acquisition.

Theoretical ideas of the symposium were presented as a written manifesto and diagnosis of the reality of that time. The manifesto emphasized the scale of overproduction and waste resulting from an increase in the number of consumed goods, confirming the rule that an increase in demand leads to an increase in production and exploitation. Consequently, the authors of this text foregrounded mental and psychological changes, the lack of balance between the "civilization"—understood by them as the style of living, and the fact that industrialization may have meant the development, and also self-destruction founded on the belief in the perpetual transformation of the world. In the authors' views, such reflection could have been encouraged by the fascination with technological advancement and progress, for example, engineering and the development of engines.

Diagnoses aim to find solutions. The authors of the text argued that while searching for answers to that problem, the status quo in science is insufficient. Science can offer no more solutions to the Word's issues. The effects of the 'advancing civilization process' pose harmful threats, like the 'total extinction of humankind' (Gołkowska and Ludwiński 1972). Consequently, justified 'anti-civilization' revolts were taking place. At this point, the authors of the proclamation referred to the hippie countercultural movements, as an 'external' side of the problem, as too destructive, undermining and negating their grounds. They hypothesized that 'inner' contestants would be more appropriate, as stated by the authors 'inherent movement' of the 'civilization' itself, as an effort of a self-definition; this situation will lead to the violent conflict of 'civilizations' (Gołkowska and Ludwiński 1972). Their suggestions of the solution to this problem were:

- defining the role of science and art as means of accelerating progress and confronting each other;
- protesting the negative results of technological advancement in the group of artists and technicians;
- creating a mutual alliance.

The open-air event was organized a year after the UN UThant report was published (1969). This document—entitled 'The Man and Environmentent'—was the result of work by several academics at the UN's request. The document aimed to draw global attention to the crisis resulting from the relationship between humans and the environment that posed serious threats, such as air and water pollution, Earth contamination, waste, and shrinking resources of the natural environment. Consequently, the UN held a conference in Stockholm two years later—in 1972 (United Nations Conference on the Human Environment), the first world conference to address environmental issues. The UN conference was acknowledged and followed even by socialist countries, such as Poland and Czechoslovakia (the former Czech Republic), despite their absence at the conference. The first principle of the Conference declaration also proclaimed that: 'Man has the fundamental right to freedom, equality, and adequate conditions of life, in an environment of a quality that permits a life of dignity and well-being, and he bears a solemn responsibility to protect and improve the environment for present and future generations. In this respect, policies promoting, or perpetuating apartheid, racial segregation, discrimination, colonial and other forms of oppression, and foreign domination stand condemned and must be eliminated (Declaration of the United Conference on the Human Environment, 1972). The Stockholm Declaration, despite being categorized as a soft law rather than a binding rule of international law, had started to influence international relations. The declaration had a considerable impact on the perception of environmental issues in the Republic of Poland, where the constitutionally grounded law for the rights to the environment was proclaimed in 1975. It directly referred to the Stockholm Declaration's first principle (Habuda and Radecki 2017: 37–38). Having anticipated, to an extent, the above-mentioned facts, the creators of the Opolno-Zdrój manifesto posed the following question: 'why was addressing environmental issues in Poland at that time of such importance', claiming that it was the country 'where at that time a lot of elements of 'civilization' are still deficient' (Gołkowska and Ludwiński 1972). The following answers were given:

- Due to the low level of industrialization in Poland, it was possible to avoid the errors made by other developed countries.
- Poland had achievements in the recovery of the natural environment, particularly of water (the experiment carried out under the patronage of the UN).
- The connection with the good tradition in the humanization of the environment: for example, the involvement of Polish architects in The Charter of Athens or the Bloc and the animating of many other artistic groups.
- Other artists have already worked on that, representing different schools and approaches.

This 'semi peripheral' progress and growth (Kola 2018: 19), expressed by the authors of the open-air event, were sufficient factors to call for a change. Clearly, the ideological agenda of the open-air event, despite the opposite tendencies in the Polish economy and social attitudes of that time manifested in many walks of life, for example by strikes and revolts, demonstrated that certain changes are not appropriate. The pursuit of progress referred to in the Opolno-Zdrój event as civilisation progress, is really, following the assumptions of the creators of the manifesto regressing to the situation posing irreversible harm to the humans living on the Earth and to the Earth itself. This threat was especially visible in the West in that decade (The U Thant Report), and was spreading to the East, as demonstrated by the event in Opolno-Zdrój. Hence, the economic and industrial backwardness in Poland, as well as its low status, noticed by the participants of the event, should act as the protection against environmental destruction and a winning card in the process of environmental protection. It should also counteract, according to the author, against degeneration of social relations that, in capitalist society, prioritise and are dominated by overconsumption and environmental exploitation.

Its presence, in its different form, was found, however, as the author attempts to demonstrate, in Polish planned and socialist economy, especially in the projects of its westernisation aiming to create a better Poland under the Communist Party rules in the 1970s.Such context was used for the term exploitation by the artists and the open-air event animators and participants (as recorded in the Catalogue—see Gołkowska and Ludwiński 1972), who raised the questions of production management, including their own art. Such practices involved classifying, categorising and controlling art or science practices of that time.

Assuming, as the author suggests, that while from the perspective of neoliberalism, control aims to measure effectiveness, the planned and centrally managed economy used control for different purposes. It was taken advantage of to maintain the regime, through censorship and eliminating undesirable voices and events, by recording and giving permission. Too individualistic expressions were not welcome, as potentially criticising, and it was the communist party in power, and was the highest elites' role to interpret and revise the reality, not the citizens.

The manifest demonstrates the beginnings of decolonising approach, as well as that of the political opposition already present in socialist Poland. The manifest proclaimed striving for release before experiencing further entanglements, some of which were already present like the sphere of Soviet influence, others were present on entering capitalism, yet under the mask of the socialist economy.

Note worthily, the author needs to indicate an interpretation gap here. The criticism of the West was well received and welcome in the East. Therefore, any anti-imperialist, anti-western, voices or moves were in line with the Party, Eastern Bloc Countries ideology, as well as the direction of the Cold War. Such ambiguity cannot be underestimated, to fully understand decolonisation entanglement in the Polish People's Republic. Therefore, this poses a question addressing who exactly and why was behind decolonisation and what it aimed to achieve and what it aimed to lead to. However, those entanglements were initiated in Poland of the 1970s, when Polish communist authorities took loans from western banks, managing this money not in the style of capitalist investors, but misused by party apparatchiks, which resulted in an internal economic crisis. This deepened the public dismay and resulted in the imposition of martial law in 1981.

The author poses a question of whether this decolonisation, present against all odds, in the manifesto, in its own, glocal cultural way, could have led the animators to the degrowth approach.

Degrowth could have been manifested by, as the author believes and remembers, what was expressed by the authors of the Manifesto addressing the delay, referring to the economy of deficit and failure in a colloquial discourse (see Kaliński 2021). No inventory announcements were a typical sign of the Polish economy of that time as well as growing numbers of frustrated shop customers. In addition, the inventory here means the necessities, not luxuries like new washing machine models. The lack of inventory triggering the anger makes the author cautious about her interpretations, whether we are dealing here with the anticipation of degrowth, or rather, which the author tends to believe, Decolonisation.

Some elements of degrowth could have been manifested in the parts addressing destructive forces of consumerism and overproduction. In the author's opinion, there is anticipation or intuition of other directions of development, presented by the creators of the open-air event in Opolno-Zdrój. Specifically, addressing the question of production aiming not to increase the level of consumption. Adding value to 'the delay' of socialist Poland would be a premonition that production intensification not only satisfies the demand, but also artificially drives it.

At the same time, the intensification of production introduced considerable social stratification, which was worthwhile, considering the society was defined by the socialist and communist rule as equal. Even if the author maintains her point concerning the anticipation of degrowth in Opolno-Zdrój, it would have been the first step in that direction.

The manifesto does not propose any alternatives to aspirations seeking development understood as growth. Neither it indicates a specific solution to tackling the problem of the deficit economy, reflecting the desire to possess and satisfy hunger in a biological sense, as well as, factually, which happened in The Polish People's Republic satisfying desires for pure consumption. The author believes that the seeds of Degrowth thinking could have been the intuition of the creators of the Manifesto, expressed in the earlier cited points, specifically, the statements addressing environmental protection by sustainable spatial design aiming at preserving the natural environment in compliance with the Athens Charter. This could have been done by means of artistic practices raising the awareness and sensitivity to the relations between humans and nature.

Another clue is inviting the local inhabitants and the mine authorities and workers to participate in the open-air event. It was the time when and place where industrialisation and coal mining were exerting a significant impact both on the local communities experiencing these phenomena and the landscape itself changing its character to an industrial one. Especially, considering the etymology of Opolno-Zdrój, 'Zdrój' means the area rich in natural resources used in a natural spa treatment. On the one hand, the region was rich in natural resources, such as mineral water and therapeutic mud used for naturotherapy and brown coal on the other hand. Both resources were discovered at the same time, in the 19th Century, but the biggest changes took place after World War Two, when open cast coal mining was applied, and the building of the mining complex power station was started in 1959. This caused irreversible changes to hydrogeological structures resulting in the disappearance of natural springs sources, while burning coal in Turow Power station in Bogatynia, located in the vicinity of Opolno-Zdrój, contributed to an increase in air pollution in the region.

Such exploitation of the resources on a large scale led to the considerable transformation of the region into a surrounded by slag heaps and gigantic mining machinery and as big as 10 × 6 square kilometres post open cast crater.

The Manifest of Opolno-Zdrój, proclaims objectives and aspirations as well as the specific glocal insight into the surrounding worlds - the closer one and the one on the other side of that reality, spreading unevenly on both sides. It demonstrates the understanding of the ideas flowing from the outside, namely, the West. It also demonstrates the interesting, as the author thinks, proof of operating these ideas juxtaposing other and own backgrounds.

The open-air event, due to its interdisciplinarity, heteroglossia, multiperspective and artistic ferment shows its readiness to turn the way of thinking. The author believes the Opolno-Zdrój event to have proved that it is possible to find glocal means of communication and to discuss and solve broadly existing issues resulting from the crises, by redefining decolonisation and degrowth, even if these terms were not directly used in the very event.

However, articulating any environmental concerns and those resulting from destructive human practices (UThant Report former, and The United Nations Conference on the Human Environment held in Stockholm later), could have resulted in interlocking the discourses on both sides of the Cold War-dominated world.

Similar concerns were articulated by Polish art addressing ecological issues. The concerns for the Earth, the human condition, and deepening shifts in the perception of human beings not as a constructive creator, were present in Opolno-Zdrój, specifically articulated from the perspective of a socialist country. However, it was the voice muffled by the East, and therefore probably neither heard nor recognised further, for example, in the West.

3.5 Discussion

Degrowth and decolonisation concepts have existed for decades, but they are hard to define unequivocally. Degrowth was first defined in 1972 (see D'Alisa et al. 2020), while decolonisation seems to have been coined by Moritz Julius Bonn, a German scholar in 1932 (Wesseling 1987).

The most popular term describing transformations brought by decolonising movements was 'the wind of change'. The metaphor was used by Harold Macmillan, the British Prime Minister, in the 1960s to describe the situation in Africa (Betts 2012: 24). Decolonisation was referred also to as 'eclipse' or 'collapse' and 'dissolution' (Betts 2012: 24). All these terms convey the geopolitical changes occurring in the world and reflect their character and strength. They mainly argued against the coloniser, aiming to challenge the imposed order. The term 'dissolution' emphasises the interesting hidden meaning of decolonisation. The prefix "dis" carries the meaning of destruction, while 'solution' indicates a new beginning. Yet, the new beginning may not always seem to follow the preconception of the anticipated change. Dissolution describes the situation driven by the 'force of inevitability' (Betts 2012: 24). "Dis-solution" serves as the frame of reference to understand degrowth and decolonisation in this sense that "dis-solution" reveals the character of degrowth and decolonisation, their driving forces and directions.

Due to their complex history and applications of degrowth and decolonisation, the author suggests, considering these concepts beyond their semantic context, and, specifically, adopting the perspectives of anthropology and ethnography, as categories falling into multi-sited conditions (see Marcus 2011).

Following multi-sited ethnographic research from the late 20th Century, multi-sited categories (see Marcus 1995) can be described as 'hybrid, displaced and recombined' (Marcus 2011: 19). The source of George E. Marcus' proposal on the multi-sited approach involves analysis of the complex phenomenon, usually based on supra-local connections. The disappearance of relatively isolated communities brings attention to the intersection of divisions between what is local and what is global (or glocal) and their mutual relations. This approach encourages the study from the perspective of a specific context, focusing primarily on a problem (often of a global nature) and its dependencies in several locations. These locations can have different characteristics: territorial, social, virtual, and ideological, to name a few (Wójcikowska 2015: 96–97).

For the author, it is important to suggest, that the meanings of degrowth and decolonisation partly overlap. They share countercultural thinking demanding transformations and fair equality in the fields of politics, culture and ecology. For this reason, the author believes that these concepts are seen as correlated. They operate in a special relationship, which in the earlier mentioned context of human turn inspired by Hastrup, the author calls a specific Decolonization Degrowth Turn. The 'turn' compels one to review the construction of the social world transcending both binarity and oppositionality. The division refers to the mentioned already a few times the 'East–West' opposition, classifying it further to the 'First', 'Second; and 'Third' world countries; 'The Global North' and 'The Global South'.

However, what the author would like to highlight is that, from her perspective, the demarcation lines alongside latitudes and longitudes do not change the logic of the polarisation process. They separate the spheres of influence and values into mutual oppositions, stressing, at the same time, their mutual isolations resulting from such oppositions. Such configuration may give an impression of symmetry. As the author believes, it still, is a misleading one, as exercising power in different contexts, prioritises opportunities to oppress diverse groups by those in power, which are diverse too. Ethics or ways of alternative resistance or subversive tactics (for instance artistic, or intellectuals' discussions) could minimise the pressure resulting from being victimised, and therefore, there is a priority for the oppressed. That is their 'weapon', thus rather a moral power than political 'hard'. In the author's view, such asymmetry is typical of not only the mutual relations between large regions but also their configurations at their internal levels, full of individual contradictions and dependencies.

As the author attempts to illustrate, such ambiguities are typical of the post-dependency studies, and therefore they support the Degrowth based approaches. The demand here is, as the author further emphasizes, to stop the course of an arbitrary process and open the doors to new alternative solutions. Hence, the decolonizing of the mind is highly desirable in the process and addressed by the post-dependency studies.

3.6 Conclusion

The paper aimed to present a way of applying the postulates of decolonisation of knowledge in the context of post-dependence studies. Auto-critical reflection on social, historical, cultural or geopolitical factors is paramount here. An example of the beginnings of such thinking in Poland and chosen by the author for interpretation is the case of artivist practices developing in the 1970s in the Polish People's Republic.

The specific glocality of the open-air event from Opolno-Zdrój, as well as the individualistic judgments exercised by artists and curators, emphasized the polemical style of the dialogue criticizing the existing patterns of development schemes and reflecting the worldwide concern, following the ideas proposed by the UN report, and criticizing capitalism. The hosts and guests of the artivist event were against the practices that enhanced environmentally destructive development; however, they were not explicitly directed against the politics behind this concept (either socialist or capitalist). Their criticism, as the author believes, was initiating the concepts of how to redirect or impede the development, eliminating its harmful effects, not its roots.

The point was to answer the question of how to design the future to avoid the pollution of the natural environment (Gołkowska and Ludwiński 1972). It was the epoch of technological advancement resulting in advanced production, even in socialist Poland. The industrialization of the 1970s enhanced the internal migration from the countryside to industrial regions across the country. The immediate vicinity of mines, power plants, the industrial landscape of the Opolno-Zdrój region, and the specific character of the open-air program inspired artists to create ecological artistic statements and discuss the relationships between humans and Nature—in different places of that town. The following open-air events were organized consecutively in the next years: the second in 1972, the third in 1973, and the last in 1974, in the same town, featuring: 'Science and art in the process of protecting the human vision zone', 'Plastic painting of water paintings' and 'Artist, and the land 400,000 kms away'. Even though the dialogues and symposium ap-proached the event from so many perspectives (represented by artists, scientists, and curators), it did not reach a consensus regarding establishing a stable pattern of thinking aiming to formalize and shape opinions on a large scale. On the other hand, however, the author proposes to interpret these as establishing of a stable initiative, and she further insists on treating these open-air events endeavors like a permanent work on the idea of degrowth, decolonized approaches, and imagination. From another perspective, critics, artists, and intellectuals participating in the event posing the questions, constantly had to reshape their own visions to push them further and shape the future alternatives, discourse, ideas, and works of art. The organizers of events tried to work on the change, decolonizing their own approaches.

Such ambition was nevertheless hard to nurture due to the centralized control of the government that had never allowed or encouraged the freedom of speech.

It seems there were discrepancies between the awareness of environmental threats and overproduction between the Polish 'ordinary' public suffering from the economic regime and that of the artivists and scientists whose representatives were present at the event. The language of intellectuals rendered the interstitial character of being 'between': on the one part influenced by 'western' terminology coined to describe new threats, also due to free access to new documents published by the UN, but on the other hand, the creators of the plain air meetings have found ways to escape from the 'socialist' style of communication. Besides, the Communist Party carefully followed and analyzed global threats, aiming to sustain socialism and its control over society.

Decolonisation Degrowth Turn originated from complexity, consisting of the interplay of diverse reflections resulting from the art and research of that period in the contexts of the 'West' and 'East'. Such practices opened eyes to discovering new ideas, concepts, and meanings, and encourage wider interpretations. They have a status of a sine qua non-factor to understand the turning point of reflection and discourse at which the change is made or implemented in practice. The glocal case discussed in this chapter, shows what is hidden in the space assessed by the Iron Curtain. Winston Churchill used the term 'shadow', introducing the Iron Curtain notion. In the shadow, "it was possible not to look too closely, permitted even to look away" (Grzechnik 2019: 2). According to Marta Grzechnik, 'the metaphor of the shadow of the iron curtain puts accent elsewhere: on ignoring Eastern Europe—not necessarily out of the ill will, but because mental maps of most Western Europeans (and Northern Americans) do not include Eastern Europe at all' (Grzechnik 2019: 2).

This is undoubtfully true, but the author of this paper attempts to show, at the same time, that this "shadow" contextualized the development of the artivist movement in socialist Poland of the 1970s. The movement created its own postulates and manifestos, designing new lines of reasoning and interpretations, playing on the meanings addressing its own topics, feeding on the information flowing from the West, as understood by the intellectualists involved in the Opolno-Zdrój event.

Entering the context of hegemony, rejecting, or accepting it, encourages the academic discussion where all the terms, such as decolonization, degrowth and post-dependency participate, projecting their voices in search of the new.

Acknowledgements This chapter is partially related to the ongoing project COST Action 19129 "Decolonizing development: research, teaching", practice, and to the other ongoing resaerch project "Decolonizing Degrowth—beyond the Eurocentric Western nature of conceptualizations and movements" at University of Łódź, Poland.

References

Betts RF (2012) Decolonization a brief history of the word. In: Bogaerts E, Raben R (ed) Beyond empire and nation: the decolonization of african and asian societies, 1930–1970s, vol 244. Verhandelingen van het Koninklijk Instituut voor Taal-, Land- en Volkenkunde. Brill, pp 23–38

Broadbent A (2017) It will take critical, thorough scrutiny to truly decolonise knowledge. The Conversation. https://theconversation.com/it-will-take-critical-thorough-scrutiny-to-truly-decolonise-knowledge-78477

Buchowski M (2006) The specter of orientalism in Europe: from exotic other to stigmatized brother. Anthropol Q 79(3):463–482

Budzyńska-Daca A, Rusin-Dybalska R (ed) (2022) Dyskursy polityczne w Polsce i Czechach po roku 1989. Wydawnictwa UW, Warszawa

Copik I (2019) Na zakręcie. Krajobraz polskiej transformacji w filmach przełomu lat 80. i 90. Kwartalnik Filmowy 105–106:23–38. https://doi.org/10.36744/kf.33

D'Alisa G, Demaria F, Kallis G (2020) Degrowth: słownik nowej ery. LangeL, Łódź

Golańska D, Kronónberg AK (2022) Creative practice for sustainability: A new materialist perspective on artivist production of eco-sensitive knowledges 1. Int J Educ Art 16(3):303–318. https://doi.org/10.1386/eta_00035_1
Gołkowska W, Ludwiński J (ed) (1972) Plener Ziemia Zgorzelecka—1971, Opolno-Zdrój, lipiec 1971: nauka i sztuka w procesie ochrony naturalnego środowiska człowieka. Wrocławska Drukarnia Kartograficzna, Wrocław
Gosk H (2012) Tożsamościotwórcze aspekty polonistycznych studiów postzależnościowych. Teksty Drugie 3(135):51–62
Grzechnik M (2019) The missing second world: on Poland and postcolonial studies. Int J Postcolonial Stud 21(7):998–1014. https://history.fas.harvard.edu/files/history/files/grzechnik_wigh_global_history_seminar.pdf
Habuda A, Radecki R (2017) Ochrona przyrody w systemach prawnych ochrony środowiska Polski, Czech i Słowacji. Difin SA, Warszawa
Harding S (2004) Introduction: standpoint theory as a site of political, philosophic and scientific debate. In: Harding S (ed) The feminist standpoint theory reader: intellectual and political controversies. Routledge, New York, pp 1–15
Hastrup K (2008) Droga do antropologii. Między doświadczeniem a teorią. Wydawnictwo UJ, Kraków
Hairong Y (2007) Positions without Identity: an interview with Gayatri Chakravorty Spivak. Positions 15(2):429–448. https://doi.org/10.1215/10679847-2006-036
Hira S (2017) Decolonizing knowledge production. In: Peters MA (ed) Encyclopedia of educational philosophy and theory. Springer, Singapore, pp 375–382
Howarth D (2008) Dyskurs. Oficyna Naukowa, Warszawa
Jabłońska B (2009) O polskim dyskursie politycznym na tematy europejskie. Debata nicejsko-konstytucyjna w prasie codziennej. Wydawnictwo UJ, Kraków
Kalicki J, Orłowska W, Kowalska E, Ziarkiewicz R, Fedorowicz J (2008) Awangarda w plenerze: Osieki i Łazy 1963–1981. Polska awangarda II połowy XX wieku w kolekcji Muzeum w Koszalinie/Avant-garde in Plein-air: Osieki and Łazy 1963–1981.Polish Avant-garde of the Mid-20th Century in the Collection of the Museum of Koszalin, Muzeum w Koszalinie, Koszalin
Kaliński J (2021) Kryzys gospodarczy w latach 1979–1982 zwiastunem upadku PRL. Optimum Econ Stud 1(103):3–23. https://doi.org/10.15290/oes.2021.01.103.01
Kastner J, Wallis B (2010) Land and environmental art. Phaidon Press Limited
Khondker H (2004) Glocalization as globalization: evolution of a sociological concept. Bangladesh e-J Sociol 1(2). https://web.archive.org/web/20051210071702; http://www.bangladeshsociology.org/Habib%20-%20Glocalization.htm
Kola AF (2018) Socjalistyczny postkolonializm. Rekonsolidacja pamięci. Wydawnictwo Naukowe UMK, Toruń
Kołodziejczyk D (2014) Wprowadzenie. In: Gosk H, Kołodziejczyk D (ed) Historie. Studia postzależnościowe w perspektywie porównawczej. Universitas, Kraków, pp 9–19
Kościk E, Głowiński T (ed) (2007) Gospodarka i społeczeństwo w czasach PRL-u (1944–1989), Gajt Wydawnictwo, Wrocław
Krukowska A, Rzerzycha-Myśliwy M (ed) (2021) Odkrywki. Kropla wiedzy o plenerze Ziemia Zgorzelecka 1971–2021. Muzeum Współczesne we Wrocławiu, Wrocław. https://muzeumwspolczesne.pl/mww/wp-content/uploads/2021/07/Odkrywki-Kropla-wiedzy-o-plenerze-Ziemia-Zgorzelecka.pdf
Laska AS (2011) Discourse as a category of analysis in political science. Athenaeum Pol Polit Sci Stud 30:31–49. https://doi.org/10.15804/athena.2011.30.02
Latouche S (2020) Dekolonizacja wyobraźni. In: D'Alisa G, Demaria F, Kallis G (ed) Degrowth: słownik nowej ery. LangeL, Łódź, pp 116–120
Lees N (2020) The Brandt Line after forty years: the more North-South relations change, the more they stay the same? Rev Int Stud 47:85–106. https://doi.org/10.1017/S026021052000039X.ISSN0260-2105

Leśniewska-Zagrodzka Anna (2006) Puławy 66: I Sympozjum Artystów Plastyków i Naukowców, 2–23 sierpnia 1966/First Symposium of Visual Artists and Scientists, August 2–23, 1966. Wydawnictwo Drukarnia L-Print, Lublin

Marcus GE (1995) Ethnography in/of the world system: the emergence of multi sited ethnography. Annu Rev Anthropol 24(1):95–117

Marcus GE (2011) Multi-sited ethnography: five or six things i know about it now. In: Coleman S, von Hellermann P (ed) Multi-Sited ethnography: problems and possibilities in the translocation of research methods. Routledge, pp 16–34

Müller M (2020) W poszukiwaniu Globalnego Wschodu: myślenie między Północą a Południem. Praktyka Teoretyczna 3(37):157–186. https://doi.org/10.14746/prt2020.3.8

Nowicka-Franczak M (2017) Ucieczka w postkolonialność. Pułapki wschodnioeuropejskiego dyskursu teorii postkolonialnej. Zeszyty Naukowe UJ. Prace Etnograficzne 45(4):403–414. https://doi.org/10.4467/22999558.PE.17.020.8679

Nycz R (2011) "Nie leczony, chroniczny pogłos". Trzy uwagi o polskim dyskursie postzależnościowym. In: Nycz R (ed) Kultura po przejściach, osoby z przeszłością. Polski dyskurs postzależnościowy—konteksty i perspektywy badawcze. Universitas, Kraków, pp 7–14

Out in the fresh air? Early ecological projects during Artistic Plein-Airs in Poland. http://www.fundacjaarton.pl/?ID=1039

Pal SK (2005) Lexicon on geography of development. Concept Publishing Company, New Dehli

Saifullayeu A (2015) Kolonializm „po naszemu" czyli o potrzebie zachowania ostrożności w używaniu metodologii postkolonialnej do badań nad Europą Wschodnią. Polemika z Mykołą Riabczukiem. Nowy Prometeusz 7:77–92

Sasanka P (2011) Polska Gierka—dekada przerwana czy zmarnowana? In: Polski wiek XX, vol 4. Bellona i Muzeum Historii Polski, Warszawa, pp 7–34. https://depot.ceon.pl/handle/123456789/1350

Solarz M (2009) Trzeci Świat. Zarys biografii pojęcia. Wydawnictwo UW, Warszawa

Sztompka P (2005) Socjologia zmian społecznych. Znak, Kraków

Thompson EM (2000) Trubadurzy Imperium. Literatura rosyjska i kolonializm. Universitas, Kraków

Treliński J, Pierzgalski A (2018) Dokumentacja działań podczas plenerów w Osiekach.Sztuka i Dokumentacja 18/Art and Documentation 18. https://doi.org/10.32020/ARTandDOC/18/2018/18

Wallas T (2010) Transformacja systemowa w Polsce i jej konsekwencje. Przyczynek do oceny. In: Wallas T (ed) Między historią, politologią a medioznawstwem. Wybór problemów. Wydawnictwo Naukowe WNP UAM, Poznań, pp 361–374

Wesseling H (1987) Towards a history of decolonization. Itinerario 11(2):95–106. https://doi.org/10.1017/S0165115300015473

Winkler A (2019) Czy za Gierka naprawdę żyło się tak dobrze? Koszty życia i zarobki Polaków w latach 70., wielkahistoria.pl

Wójcikowska K (2015) Wykorzystanie etnografii wielomiejscowej w badaniach procesów migracji edukacyjnej wśród mniejszości litewskiej w Polsce. Rocznik Lubuski 41(1):93–106

Wood DA (2018) Epistemic decolonization: a critical investigation into the anticolonial politics of knowledge. Palgrave Macmillan

Worłowska M, Krawczyk N, Szymczak B (2019) Klęska urodzaju. Początki sztuki ekologicznie zaangażowanej w Polsce. https://galeriaopole.pl/wystawa-kleska-urodzaju-/

https://plus.gloswielkopolski.pl/humor-w-prlu-jak-nas-smieszyly-czasy-slusznie-minione-czyli-humor-czasow-cenzury/ar/c11-14290473

https://isap.sejm.gov.pl/isap.nsf/DocDetails.xsp?id=WDU19690130094

https://worldservice.org/stockholm.html

https://artmuseum.pl/pl/archiwum/archiwum-eustachego-kossakowskiego/16?read=all;

https://www.money.pl/gospodarka/w-tym-miescie-bieda-przez-lata-miala-twarz-kobiety-nie-palily-opon-nie-upominaly-sie-o-swoje-6357107758978689a.html

https://www.history.com/topics/1970s/1970s-1

http://www.journal.doc.art.pl/pdf6/SID_6_all.pdf
https://www.un.org/en/conferences/environment/stockholm1972

Inga Barbara Kuźma associate professor at the University of Łódź, the Institute of Ethnology and Anthropology of Culture, and at the Centre of Social Innovations; the member of SIEF, F.E.R Eurethno. The expert of FEANTSA (the Housing Rights Cluster). Her research interests: herstory, social exclusion and inclusion, public and practical anthropology institutional violence and its hidden curriculum, and spaces of aid. She is also the co-founder of the Interdisciplinary Seminar Gender. She is interested in urban social activism, i.e., she had taken part in the Municipal Council for Solving the Problem of Homelessness in Łódź and she is a member of herstoric Foundation Łódź Women's Trail.

Chapter 4
A New Wave of Civic Activism

The Case of Social Movement Against the Construction of the Amulsar Gold Mine in Armenia

Adam Pomieciński

Abstract This chapter focuses on the radical activities and actions of the environmental movement related to the exploitation of gold in Amulsar, as well as global growth issues in the context of anti-capitalist logic and degrowth revision. The Amulsar gold mine is located 170 kilometers south of the capital of the Republic of Armenia, Yerevan. It was created by the international concern Lydian International. This investment caused a lot of controversy among both the inhabitants of Armenia and environmentalists. Activists and independent experts pointed to the risks associated with the construction of a gold mine, including irreversible damage to the natural environment, destruction of the entire water system of Armenia and destruction of the cultural heritage of the region. In 2018, the construction of the gold mine was completed and mining began. This caused the indigenous environmental movement to take radical steps to block activities in the Amulsar area. Protests continue to this day. The example of the Amulsar gold mine in Armenia shows the consequences of global growth for the peripheral countries.

Keywords Radical movement · Social movement · Armenia · Amulsar · Gold mine · Political economy · Limits to growth

4.1 Introduction

Armenia, like other countries of the South Caucasus, grapple not only with ethnic and territorial conflicts (Pomieciński 2022: 69–88), but also with many other problems. Among them, one of the most serious are environmental problems that affect the lives of the inhabitants of this region. The researchers, among others, note the exploitation of transboundary waters, which very often become the target of military operations in conflict zones, and their control gives an advantage to the parties on the battlefield (Veliyev et al. 2020). Another challenge related to the environmental issues is the increase of emission in harmful substances that are polluting the air. Air pollution,

A. Pomieciński (✉)
Institute of Anthropology and Ethnology, Adam Mickiewicz University Poznań, Poznań, Poland
e-mail: adpom@amu.edu.pl

M. K. De Santo and S. E. Domptail (eds.), *Degrowth Decolonization and Development*,
https://doi.org/10.1007/978-3-031-25945-6_4

for example, in Yerevan, the capital of Armenia, often exceeds the rates of cities such as Moscow or Istanbul (Martirosyan and Khachatryan 2020). In addition, there is the illegal exploitation of wood and forest devastation (Mejlumyan 2019a, b), as well as lack of proper waste management, which is associated with the terrible condition of landfills and fragmented efforts to segregate them (Muradyan 2021). However, many argue that the greatest challenge for Armenia today is to prevent the negative impact of mining on the environmental and social situation (see, among others the EcoLur Press Club Report 2019). It is mining that has the most detrimental effect on a number of social and ecological spheres. In this context, the Amulsar gold mine construction project has been recognized as the most serious environmental problem for Armenia in recent years.

Article focuses on a new wave of civic activism and the latest social movement against the construction of the Amulsar gold mine in Armenia. This "Golden Project", raises a number of issues. The processes are characteristic of universal capitalismand are based on territorial expansion, imposing its own rules on areas often regarded as peripheral. The desire to profit at the expense of overexploitation of the environment and people underlies global forms of market imperialism and is materialized in the Amulsar gold mine project, for which a multinational mining company is responsible.

As Armine Ishkanian argued, the civil initiatives in Armenia often deal with very specific and narrowly focused issues, such as saving a waterfall or a public park. However, their appearance expresses much broader concerns, which are related to corruption issues, the lack of the rule of law, the lack of democracy, the rise of oligarchic capitalism, the failure of political elites to solve the problems of "ordinary" Armenian citizens, etc. (Ishkanian 2015). All these fears and feelings are also portrayed in the social movement against the construction of the Amulsar gold mine. Ecologists, activists and local residents have taken a confrontation against the international mining company and the corrupt policy that continues to leave its mark on the post-Soviet mining industry of today's Armenia.

Public social movement against the construction of the Amulsar gold mine is part of the efforts of the global degrowth movement to combat the climate crisis, which has both global and local dimensions. In other words, the actions of local activists, representatives of environmental movements and the inhabitants themselves are focused on opposing and stopping the construction of the Amulsar gold mine in Armenia. They speak publicly about the limits of global growth and the exploitation of natural resources by transnational corporations. The aspirations of the opponents of the Amulsar gold mine are analyzed in the context of the basic assumptions of the degrowth idea (e.g. limiting production, limiting the activities of the mining sector, protection of the environment and local culture, pluralism) and related practices (activism and social protests, radical actions, generating alternative knowledge). The movement against the Amulsar mine indicates significant problems related to the environment (environmental degradation, over-exploitation) and the cultivation of the region (destruction of the cultural heritage of the city of Jermuk, considered a tourist and spa town, in the vicinity of which the mine is located).

The main purpose of this article is to analyze the social activities of the opponents of the Amulsar gold mine in Armenia, which are the result of tensions arising around two contradictory ideas: unlimited development of the world economy and alternative concepts undermining economic growth as a tool of political and economic goal. The case of Armenia shows that the post-Soviet states experience dependence both on regional powers (in the case of Armenia, these are primarily Russia, but also Azerbaijan and Turkey, focused on military expansion), as well as international corporations driven by unlimited economic growth and exploiting valuable natural resources that are part of the heritage of local communities. This causes new dependencies for them that take place in the areas of politics and economics. Thus, new forms of civic activism in the face of the crisis caused by the construction of Amulsar are today a huge challenge for shaping new ways of thinking and views that undermine the validity of the "growth economy" in one of the poorest post-Soviet country.

4.2 Ecological Movements in Post-soviet Armenia

At the time when Armenia was part of the Soviet Union, all grassroots social movements were effectively suppressed. The communist authorities considered them dangerous and threatening the stability of the system (Fitzpatric 1999: 7–9). Public activities were organized exclusively by the state and nothing could have happened without the knowledge of the communist party. As indicated by Elena Zdravomyslova and Viktor Voronkov, social movements (mainly dissident and human rights) began to appear in the USSR in the 1960s and consisted mainly of liberal intelligentsia. In the late 1960s and early 1970s, Soviet authorities quelled opposition, and many dissidents either emigrated or were imprisoned (Zdravomyslova and Voronkov 2002: 97–117).

The first protest in the Armenian Soviet Socialist Republic, which were not controlled by the central authorities and the communist party, took place in April 1965. Thus protestwas among the greatest in the Soviet Russia era. However, they did not concern ecology, but the issue of recognition and commemoration by the authorities in Yerevan and Moscow of the Armenian genocide that took place on their Anatolian ancestors in the Ottoman Empire in the years 1915–1917. The issue of the Armenian genocide could not be raised in public discussions so far, as ethnic and national problems were effectively silenced by the party leadership. Following this brief political thaw, the authorities took a more conservative course of their rule, causing grassroots movements and social initiatives to be essentially blocked for many years.

The situation changed only in the 1980s, when Mikhail Gorbachev came to power and started the policy of perestroika. Social changes affected almost all spheres of activity of the Soviet state: from the reduction of censorship and democratization of public sphere to market mechanisms implementation projects.

This new policy provided opportunities for public activism. The general situation of the USSR, which was beginning to decline, also contributed to this. It was in the mid-1980s that the labor, national, and environmental movements became active. In

Armenia, the first environmental movements were initiated in 1986–1988. Activists demanded the closure of the Nairit Chemical Plant and the Mecamor Nuclear Power Plant. Air pollution in Yerevan and other Armenian cities posed a serious threat to the inhabitants. In the 1980s, environmental changes as a result of industrialization and urbanization and the increase in various diseases among the population of the republic began to be noticed more and more. They were manifested especially in the Ararat valley around Yerevan, where the greatest concentration of industry and population took place. One of the manifestations of the social reaction was the open letter of 350 Armenian intellectuals against environmental poisoning, which they addressed in March 1986 to the Soviet leader Mikhail Gorbachev. The leadership of the Armenian SSR also decided to take up these issues, which was reflected in the announcement by the First Secretary of the Communist Party of Armenia, Karen Demirchian, to reduce pollution at the main chemical plants in Yerevan and Kirovakan. The slowness of the authorities in solving this type of problems and the disclosure of the actual size of the pollution led to demonstrations of the environmental movement in Yerevan in the second half of 1987 (Nieczuja-Ostrowski 2016: 65).

In mid-1988, the Armenian ecological movement began to change rapidly, including other issues in its postulates. For example, there were also demands for more civil rights for Armenians, political reforms, and ultimately for the independence of Armenia itself. Thus, the problems of ecology and the environment gained additional political and national meanings. For Armenians, environmental pollution began to be treated in terms of ecological genocide, just as discrimination against Armenians in Nagorno-Karabakh was equated with cultural genocide. The Armenians linked these two issues together. Thus, the category of ethnic genocide was expanded and included problems related to the environment. This change made the Armenian ecological movement rapidly evolve towards a national movement with clear political aspirations (Geukjian 2007: 233). The effect of this change was the transformation of environmental movements into a national political movement, which became known as the Karabakh movement (Ordyan and Pomieciński 2021: 193–212). It later contributed to the collapse of the Soviet Union and the declaration of Armenia's independence in 1991.

For many years, environmental issues in Armenia have been relegated to the background. Independent Armenia from the very beginning of its existence began to deal with strong political upheavals and the war over Nagorno-Karabakh. In addition, there were also serious economic and social problems. This caused environmental issues to remain temporarily postponed until the late 2000s. The revival of environmental initiatives only came in 2007, when a citizens' initiative to stop copper and molybdenum mining in the village of Teghut in the north of Armenia was launched. The protests of activists, mainly young people, were directed against both international development agencies that supported the Teghut open-pit mine project, and the Armenian government, which allowed the exploitation of the deposits (Ishkanian 2016). It was actually the beginning of a new activism against the over-exploitation of Armenia's natural resources. Although the activists did not stop the opening of the Teghut mine, they managed to challenge the interests of international corporations, investment funds and oligarchs (Ishkanian 2016), and to publicize in Armenian society the issue of nature conservation treated as public property. The protesters also drew attention to many glaring irregularities and revealed a number of negative effects of mining activities, including destruction of many hectares of rich forest, contamination of groundwater and local rivers, or failure to comply with agreed environmental standards.

Parallel to the initiatives related to the operation of the Teghut open-pit mine, environmentalists have engaged in other protests. One of such actions was the fight to protect the Trchkan waterfall from the construction of a hydroelectric plant by the Robshin company. The Trchkan Waterfall, on a tributary of the Chichkhan River, located on the border of the Lori and Shirak regions, was part of a nature reserve and park. Nevertheless, to the surprise of many people, the Armenian government decided to extend the power plant's concession. In early September 2011, environmentalists decided to protest. Thanks to social media, the group of protesters grew and received support from journalists and residents of nearby towns.Supporters of the movement not only from Yerevan, but also from Gyumri, Vanadzor, Etchmiatsin and Armavir gathered in a tent camp set up by activists near Trchkan waterfall. On November 3, environmentalists were successful when then Prime Minister, Tigran Sargsyan, called for the end of the construction of the power plant. The "Protect Trchkan Waterfall" civic initiative has been hailed as a landmark victory in Armenia, proof that civic activism and a general social movement for environmental protection can effectively reduce economic activity that depletes natural resources and destabilizes the ecosystem (Avanessian 2011). Similar social movements were initiated on other environmental issues.

In early 2012, Armenian environmentalists became involved in the protection of Mashtots Park in Yerevan. The city authorities approved a new plan to build shops and boutiques at the expense of the park's liquidation. The construction works that were approved by the city caused serious concern among many NGOs, environmentalists and the inhabitants of Yerevan themselves. That parkwas one of the few remaining public green spaces in the capital of Armenia. The occupy-style protests continued despite snow and cold (Barsoumian 2012). The protesters raised many important issues, such as the illegality of building permits, the destruction of green zones in Yerevan or the domination of the private sector over the public sector. It is also important that the mass appearances of "Save Mashtots Park"were directed against the interests of Armenian oligarchs and became a means of fighting against "old thinking based on patriarchal principles" (Saroyan 2016). The 2000s in Armenia were a time of stimulation of grassroots social movements, not only of an ecological nature. After the actions of the "Mashtots Park Movement", a wave of other social protests swept through Armenia.After the rigged presidential elections in 2013, mass student strikes began, which sparked a discussion on the rule of law. As it was written in the Armenian press: "The strikers are in a head to head confrontation with the "culture of hegemony", as described by Antonio Gramsci, that permeates all aspects of life in Armenia" (https://hetq.am/en/article/23886). Another initiative was "I'm Against", which was a reaction to the changes in the pension system made by the government in 2014 (Saroyan 2016), followed by Electric Yerevan, i.e. mass protests that took place in the summer of 2015 against the increase in energy prices (Avedissian 2015).

New social movements that have appeared in Armenia in recent years indicate the growing interest of the Armenian public opinion in such issues asdemocratization of social life, fighting corruption, limiting the political and economic influence of oligarchs and international corporations, as well as conscious protection of natural resources and the environment.In particular, eco-activist movements pointed to the serious failures of the state, which not only withdrew from the policy of protecting its natural resources at the expense of making short-term profits shared with private corporations, but repeatedly made controversial decisions favoring the destruction of the ecosystem and putting economic resources above the interests of ordinary residents. Importantly, one of the most important achievements of this movement was raising public awareness of real environmental problems in such a way that the inhabitants themselves would take their own actions to protect their environment (Veliyev et al. 2020). It was according to this scenario that the grassroots movement against the construction of the Amulsar gold mine developed.

The case of this mine more broadly reflects the practice of the mining sector in Armenia, which in some respects has not changed much since the Soviet times. For years, mining has been recognized by the governments in Armenia as an important sector of the economy. It is also the sector which accounts for more than half of the country's exports. The source of Armenia's income, however, is not shares in these mines, but mainly profits from license fees. Investors are mainly international companies from the United States, Great Britain, Canada, China, Germany and Russia.A large part of the mines is also owned by Armenian oligarchs (Ishkanian 2016). In recent years, several hundred mining concessions have been issued without any sustainable resource use programs, appropriate tax legislation, environmental protection or environmental and social impact assessments. It is worth mentioning here that Armenia has significant resources of copper, zinc, iron, lead, rhenium, cadmium, tellurium, silver and gold. According to rough estimates, there are about 613 mines in Armenia with the capacity to extract 60 types of various minerals. The high saturation of the mining sector causes that it has catastrophic consequences for the inhabitants and the environment. In such a small country as Armenia (the total area of this country is 29,800 km^2 with about 3 million inhabitants) such a situation is difficult to accept (Vardanyan 2020: 12). Hence, a strong anti-mining movement emerged as part of the ecological movement (Ishkanian 2016). Its activities entered a new phase when a new, and at the same time very controversial, project to build a gold mine in Amulsar appeared on the horizon.

4.3 Amulsar's "Golden Project"

Amulsar is one of the peaks of the Zangezur mountain range in the south of Armenia. Deposits of gold-bearing quartzite are located inside of the mountain, on the border of the regions of Vayots Dzor and Syunik within the northern ridge, north-west branch of the Zangezur range, at an altitude of 2,500–2,988 m above sea level. As a result of geological research of the Amulsar Mountains, gold resources were estimated and then given a commercial value. Gold reserves are 73,733 kg (average grade is 0.78 g/t). In addition to gold, the inside of the mountain also holds 294 367 tone of silver (the average grade is 9.29 g/t). The deposit is to be explored by a combination of Tigranes-Artavazdes and Erato opencast mines with an annual extraction capacity of 10 Mt of ore. The metals from the deposit are to be extracted by the heap leaching method, which has not been used in mining in Armenia so far (https://www.lydianarmenia.am/eng/pages/About%20the%20mine/70/).

The mine will be operated in an open manner, which means that using explosives, the raw ore will be separated from the ground, causing the emission of a huge amount of dust in the air. Thus, the wind can carry various types of heavy metals from the mine even several kilometers away (http://geopolitics.am/archives/15567) and pollute both the surrounding nature and the crops—especially fruit orchards—which are the basis of the region's agriculture and one of the main economic activities of the inhabitants of the surrounding villages. The aforementioned rock leaching method will be applied to crushed ore transported by a conveyor to the leaching heap where gold will be obtained. A dilute cyanide solution will be applied to the heap, which is necessary to extract gold from the rocks (https://www.lydianarmenia.am/eng/pages/eng/pages/amulsar-heap-leaching/124/).

The use of gold cyanization is associated with numerous threats to the environment, especially when the entire process takes place outdoors. If proper safety measures are not followed, severe consequences for people and the surrounding ecosystem can occur. While measures are taken to ensure that no hydrogen cyanide is formed, other forms of cyanide still pose a risk to living organisms. Harmful chemicals are produced during cyanization, including nitrates and thiocyanates, although their impact is much less than that of cyanide leakage. Gold mining companies must therefore adhere to strict security measures to prevent such events from occurring (https://www.netinbag.com/en/manufacturing/what-is-gold-cyanidation.html). Nevertheless, the use of cyanide in mining remains a highly controversial method due to its toxicity. One of the most recent disasters in Baia Mare, Romania in 2000, showed that the safety measures taken may not be sufficient. At this gold mine, heavy rains, ice and snow damaged the tailings dam, releasing 100,000 cubic meters of cyanide-rich waste into the watershed. 2.5 million people were cut off from drinking water, including in neighboring Hungary and Serbia. Hundreds of tons of fish were killed in the Shamos-Tisza-Danube river (https://www.europarl.europa.eu/meetdocs/2009_2014/documents/envi/dv/envi20130925_info-cyanide_/envi20130925_info-cyanide_en.pdf).

In the case of the Amulsar mine, it was assumed that it will be operated for 9 years, after which all gold reserves in the substrate will be excavated and the mine will turn into a landfill. About 122 million tonnes of ore shredding residue and 343 million tonnes of waste rock containing very harmful elements will remain (http://www.armecofront.net/lrahos/inchu-chi-kareli-shahagorcel-amulsary/). It has also been assumed that the mine closure period will be 5 years, which is inconsistent with international law stipulating that the minimum period of mine closure is 30 years. This means that for the next 25 years the obligation and the cost of closing the mine would be shifted to the state. According to some reports, the mine closure period should be longer for Amulsar and amount 65 years (Sargsyan 2019: 107–108).

The permit to build a gold mine in Amulsar was obtained by Lydian Armenia, which was officially established in 2005. At that time, all shares belonged to Lydian International Limited, which was incorporated in the offshore (so-called "tax haven") area of the British island of Jersey (http://geopolitics.am/archives/15567). In March 2021, Lydian International was dissolved and control of the Amulsar gold mine assets was transferred to Lydian Ventures of Canada. Before that, Lydian International's primary focus was on the large-scale, low-cost business driven Amulsar gold mine project.It was estimated that an average of 225,000 ounces of gold mined in Armenia per year would make Lydian one of the largest gold producers in the world (https://www.linkedin.com/company/lydian-international-limited/about/). Lydian, after submitting environmental and social impact assessments approved by the Armenian government, began construction work in 2015, while planning to fully start operating in 2018. However, independent experts as well as environmental associations and residents have questioned the entire project as potentially dangerous to the environment and people (Khachatryan 2020). Against this background, civil protests began in Armenia in order to completely block the construction of the mine in the vicinity of the golden-bearing mountain Amulsar.

4.3.1 The Fight for Amulsar

In 2009, Lydian Armenia received the first positive decision to mine gold in Amulsar from the Ministry of Environment of Armenia. From then on, the preparatory work began to progress rapidly and the company obtained a loan from the European Bank for Reconstruction and Development (EBRD) to meet its goals. In 2012, Lydian Armenia commenced the first exploration drilling in the Amulsar area. As a result, in 2012, environmentalists and activists began to organize actions against the mine. It was only the beginning of a long struggle and a dispute between the broadly understood ecological and civil movement against the construction of the Amulsar mine and the international mining corporation and the Armenian government that agreed to this investment.

In addition to environmentalists, the local population also joined the fight against the gold mine. First of all, the inhabitants of the city of Jermuk—a health resort located on a mountain slope in southern Armenia, widely known for its numerous hotels and sanatoriums, as well as bottled mineral water called Jermuk. A few years earlier, the authorities in Yerevan and Jermuk decided to transform this post-Soviet city into a resort of international standard. Critics of the mine's construction began to warn that the open-pit gold mine posed an ecological threat to the picturesque Jermuk and will cause its collapse (Abrahamyan 2012). At that time, the authorities and the mayor of Jermuk spoke out strongly against the project. A year later, after Lydian donated approximately 150,000 USD to the city's development fund, the mayor not only withdrew his support for the protesters, but openly expressed his approval of the gold mine project. In 2013–2014, Lydian donated almost 250,000 USD to the Luys Foundation, owned by the Serzh Sargsyan family, the president of Armenia in 2008–2018 (Armenia Bankwatch Network Report 2020: 17–18). Such Lydian practices among activists and ordinary residents were perceived as corrupt, as well as showing the great determination of the mining company, which tried to silence its opponents by all means.

In 2012–2014, the Ministry of Environment of Armenia issued Lydian two more positive conclusions on the assessment of the mine's environmental impact, and the Armenian government adopted a decree authorizing the transfer of endangered species of wild animals from Amulsar to another area (Armenia Bankwatch Network Report 2020: 18). In response, activists of the environmental movement and local communities tried to challenge the decisions by legal means, referring them to, inter alia, the Human Rights Defender, the EBRD Project Complaints Mechanism and the Administrative Court. Protest letters were sent to the government and the president of Armenia. The conflict began to concern fundamental issues regarding the assessment of the effects of the mine's operation and its impact on the environment and people. The expertise that the protesters had at their disposal was diametrically opposed to the conclusions used by the supporters of the mine construction. The anti-mining movement was particularly prolific in generating new knowledge about the consequences of the Amulsar "golden project".

The protesters pointed out that the threats posed by the mine are both local and global. They mainly concern Armenia's strategic water resources. The mine is located in the catchment area of the Arpa and Vorotan rivers, as well as near the Kechut and Spandaryan reservoirs, which flow into Lake Sevan—the largest alpine lake in Armenia, as well as the largest lake in the Caucasus. These waters play an important role in the hydrological economy of the entire region, as well as in the irrigation of agricultural land. The danger of water pollution and the exploitation of gold with the help of the toxic substance of cyanide pose a serious threat to the entire water system of the country and the region (Armenian Environmental Front 2014). Open-pit mining will also generate large amounts of sulfur that can get into the river network. A side effect of gold production is the release of radioactive metals, which may pose a radiation hazard to the environment and people (Ionesyan 2011). Thus, the exploitation of the mine that threatens water resources poses the future of Jermuk as a resort city and other towns whose economy is based on agriculture. In addition, the development of the mine is associated with the risk of the extinction of a number of unique species of flora and fauna. Activists also noted that Amulsar is located within a large tectonic fault that poses seismic hazards (Sanasaryan 2018). The questioning of the safe operation of the Amulsar gold mine in this discourse was based on the belief that no one can rule out the risk of environmental contamination even if all mining standards are followed. The participants of the protest also raised the issue of socio-cultural problems caused by the mine, which are directly felt by the people of Jermuk. The entire foothills and the province of Vayots Dzor have been known for centuries for viticulture and wine production. Wine traditions preserved for generations are a phenomenon of the cultural heritage of this region. Residents of settlements in the Arpa River basin are strongly opposed to the exploitation of the Amulsar as they believe that the mine will threaten their ability to grow wine and viticulture, agriculture and horticulture, leading to a socio-cultural disaster. The widespread arguments led to the fact that the wave of resistance against the construction of the Amulsar gold mine gained many social actors: from protesting farming communities determined to protect their lifestyle, ordinary people concerned about the potential contamination of drinking water supplies, to radical ecologists defending nature and habitats the most endangered species of flora and fauna (Ellis 2019). In the spring of 2016, Lydian received another, fourth permit for the extraction of gold-bearing quartzite from the Ministry of Environment of Armenia, and in the fall of that year announced his readiness to start work.In 2016, direct protests also intensified, which were organized by activists of the environmental movement and its supporters. They included Internet petition, protests in front of the offices of Ameriabank, which joined the financing of the Amulsar mine, and protests in front of the building of the Ministry of Environment of Armenia.It should also be mentioned that since 2016, a rebellion has been growing in the entire Armenian society against the government, which was finally sealed by the Velvet Revolution in 2018. It caused a change of government in Armenia, which resulted in the removal of Serzh Sargsyan and his party from power. Nikol Pashinyan became the head of the government. This political change gave new hope to the mine's opponents, who expected that now their demands would be met and that the activities of the international mining company in Amulsar would be stopped.

Nikol Pashinyan, as an opposition politician, also expressed his disagreementto the Amulsar project, as Lidian received permission from his political opponent Serzh Sargsyan. In premature elections, activists saw him as their ally (Mejlumyan 2022), and he himself supported the environmental movement and protests against the mine. After coming to power in 2018, the current Prime Minister of Armenia, Nikol Pashinyan, has promised that his government will re-examine gold mine projects. The Amulsar mine problem became one of the first major political challenges for the new Pashinyan government, forcing it to seek a compromise between grassroots activists who helped to come to power and investors who promised economic development (Jardine and Atanesian 2018). At meetings with protesters and representatives of the mining company, he announced an independent audit of the project. However, 2018 in Armenia was a year of large protests by environmentalists and citizens dissatisfied with the progressing work of the gold mine. In this way, activists gave Pashinyan a sign that the Amulsar problem must be resolved as soon as possible. In 2018, Lydian Armenia employed 1,400 people. In order to prevent the extraction of deposits from Amulsar, members of the anti-mining movement initiated a blockade of the mine in June 2018, which lasted over two years.

Under the influence of strong civil protests in the summer of 2018, the mine's development was almost completely halted. Members of the anti-mining movement blocked all roads leading to the mine. They set up small camps and temporary houses around the mountain so that the blockade could last day and night. Thus, they effectively prevented workers from entering the mine. It also sparked counter-protests by Lydian workers, who demanded that protesters be removed from the access roads to Amulsar (Jardine and Atanesian 2018). The Pashinyan government refused to remove the blockades. It also temporarily suspended Lydian's right to operate in Armenia pending an environmental audit of the Amulsar mine by an international advisory group. The consulting group hired for this audit, Earth Link and Advanced Resources Development (ELARD), published its report during a live teleconference in August 2019. Summarizing its report, ELARD concluded that the data collected by Lydian in the Amulsar study was insufficient for a mining project of the proposed size. Moreover, it was found that Lydian used these incomplete data to draw overly simplistic and unsupported conclusions. The report also concluded that the potential risk of water contamination is low and can be mitigated. In this vein, ELARD issued 16 recommendations to Lydian about mitigation measures. Following the publication of the ELARD report, the Armenian government sided with Lydian, Prime Minister Nikol Pashinyan announced in August his intention to restore operational rights. On August 19, he announced—"At this moment the mine should be exploited" and added that "Amulsar will be operating to very high environmental standards, unprecedented for Armenia" (Mejlumyan 2019b). On that day, activists began gathering at rallies to undermine Pashinyan's report and decisions. So far, environmentalists' hopes for a change in the political course of the Armenian government regarding the Amulsar gold mine have been severely strained.

The direct actions of the civic movement against the Amulsar mine escalated in August 2020, when Lydian's security personnel began dismantling camps and temporary shelters of environmentalists, who were blocking the roads to Amulsar. Activists with the help of Facebook, Instagram, Tik Tok and Twitter announced a massive mobilization of all their supporters and allies. There were clashes between the protesters and the mine's security guards and the police. As a consequence, many demonstrators were detained. Protests lasted not only in the vicinity of Amulsar, but also moved to Yerevan in front of the government buildings. Due to the threat of the COVID-19 epidemic and the ban on gatherings, the police started detaining people.In response, the protesters announced a bicycle rally. However, in this case, the police reacted quickly, arresting several people and confiscating their bicycles. Similar clashes took place several more times (Khachatryan 2020).

The situation suddenly changed in September 2020. At the end of that month, the war between Armenia and Azerbaijan over Nagorno-Karabakh (Pomieciński et al. 2022) flared up again. The blockade of Amulsar ceased, and some of the protesters became involved in the defense of Nagorno-Karabakh. As a result of the war, the border with Azerbaijan shifted, bringing it closer to Amulsar. The war moved the problem of the gold mine to the background for a while. However, the economic needs of Armenia in the post-war period increased the likelihood of the mine reopening. In 2022, the government in Yerevan made legislative changes to the mining code, enabling companies "to carry out mining with environmental impact assessments more than a year old, as long as the delay was caused by reasons that include civil disobedience" (Mejlumyan 2022). Environmentalists questioned the adopted legal solutions. Currently, non-governmental organizations are demanding the withdrawal of adopted laws, contesting amendments to the code, and trying to discredit Lydian Armenia's subsequent applications for permission to use the water necessary for gold mining. Thus, the dispute between the civic movement and the international mining company and the government of Armenia enters a new stage.

4.4 Conclusion

The conflict between the civil movement against the construction of the Amulsar gold mine is one of the longest-lasting conflicts of this type in Armenia and the entire Caucasus. Of course, its specificity is not only based on this. There are a few characteristic issues worth paying attention to. First, the anti-Amulsar movement remains part of the anti-mining movements that emerged in Armenia shortly after independence. The activists' actions are therefore part of a long history of protests against an irresponsible mining policy, which did not translate into the real development of this country, where the official unemployment rate is at the level of 15–20% (Trading Economics 2022), and many Armenians are constantly deciding for economic emigration to Europe or America. Second, the "Save Amulsar" movement has placed the mine issue in the broader context of global trends, raising anti-corporate issues and limiting global economic growth causing degradation of the natural environment and local communities' lifestyle. The so-called "self-regulation of the world market" threatens the regulation of sovereign states, especially those we used to call "developing countries". Thus, the movement against building the Amulsar mine contains many elements of post-growth thinking and acting. As Corinna Burkhart, Matthias Schmelzerand, and Nina Treu put it, "post-growth is in motion/in movements", which means that many of the latest social movements not only redefine and critique neoliberal economics, but propose alternatives and ways to move beyond "expansive modernity" (Burkhart et al. 2020: 10). This is what the protesters against the Amulsar mine seem to be saying, which aim at the welfare of all, ecology, sustainability and stopping a system based on a hegemonic economy. The applied resistance and the spread of the protest were oriented precisely at achieving these goals.

The crisis in the vicinity of Mount Amulsar created an urgent need for convergence of struggles, bringing together many different groups that felt aggrieved and dissatisfied with this course of events. Environmentalists, residents of Jermuk and other villages of the Vayots Dzor region, students and scientists, local authorities and "ordinary" citizens worried about the vision of an ecological disaster have taken joint actions to protect Amulsar. By calling themselves "Amulsar Protectors" they spread grassroots resistance practices involving individual and collective actors.Hence, the movement against the construction of the Amulsar gold mine is a movement not only fighting for the mountain, but also for the preservation of democracy, ensuring the dignity and health of all beings. Its social significance is particularly important in Armenia, a post-Soviet country that still experiences the influence of foreign powers intersecting with the interests of international capital and corporations. It calls on accountability of those who abuse their power at the expense of the environment, health and human life.

The actions of the movement against the construction of the Amulsar mine are a clear example of the development of grassroots civic initiatives in the post-Soviet areas. The new awareness of the growing problems, especially among the young generation that did not experience communism on themselves, sets new directions for social changes. The civic vote on Amulsar has become a form of grassroots politics, which at the moment is an effective political "project" blocking the authoritarian aspirations of the government and the mining corporation. The fight for Amulsar is therefore a fight against the dominant system, consolidated around the ideas of pluralism, democracy and post-materialism. The anti-gold mine movement in Amulsar remains a problem-oriented and horizontally structured civic movement seeking alliances to achieve ecological and social change, but at the same time revealing the wider mechanisms of power and corporate power. The use of civil disobedience by publicizing problems, blockades and occupying places, generating alternative knowledge in the form of diagnoses and reports forces the authorities and corporations to refrain from the decisions made and previously planned activities. The lack of clear solutions related to Amulsar means that the fight is still going on. So it is likely that activism and protests around the gold-bearing mountain will take a new form, fueled by people's anger at the lack of the rule of law and corporate exploitation.

References

Abrahamyan G (2012) Armenia: is gold mine a blessing or a curse? https://eurasianet.org/armenia-is-gold-mine-a-blessing-or-a-curse

Armenian Environmental Front (2014) Hidden dangers of Lydian International's gold mine: cyanide. https://armecofront.net/en/news/hidden-dangers-of-lydian-internationals-gold-mine-cyanide-2/

Avanessian N (2011) The saga of Trchkan. Armen Wkly. https://armenianweekly.com/2011/11/30/the-saga-of-trchkan/

Barsoumian N (2012) Battle over Mashtots Park: protesters block city plan to build on Yerevan 'Green Land'. Armen Wkly. https://armenianweekly.com/2012/02/29/battle-over-mashtots-park-protesters-block-city-plan-to-build-on-yerevan-green-land/.

Burkhart C, Schmelzer M, Treu N (eds) (2020) Degrowthin movement(s). In: Exploring pathways for transformation. Zerobooks, Winchester, Washington

Ellis G (2019) Mining versus the environment: the battle over Armenia's Amulsar gold mine. https://www.aljazeera.com/features/2019/1/31/mining-vs-the-environment-the-battle-over-armenias-amulsar-gold-mine

Fitzpatric S (1999) Everyday stalinism. In: Ordinary life in extraordinary times: Soviet Russia inthe 1930s. Oxford University Press, New York, Oxford

Geukjian O (2007) The politicization of the environmental issue in Armenia and NagornoKarabakh's nationalist movement in the South Caucasus 1985–1991. Natl Pap 35(2):233–265

Ionesyan K (2011) Amoulsar: reserves of hazardous uranium and valuable gold. https://www.ecolur.org/en/news/mining/amoulsar-reserves-of-hazardous-uranium-and-valuable-gold/3129/

Ishkanian A (2015) Self-determined citizens? A new wave of civic activism in Armenia. https://www.opendemocracy.net/en/selfdetermined-citizens-new-wave-of-civic-activism-in-armenia/

Ishkanian A (2016) Challenging the gospel of neoliberalism? Civil society opposition to mining in Armenia. http://eprints.lse.ac.uk/64900/1/Ishkanian_Challenging%20the%20gospel%20of%20neoliberalism_author_2015.pdf

Jardine B, Atanesian G (2018) Mining dispute threatens Armenia's post-revolutionary political consensus. https://eurasianet.org/mining-dispute-threatens-armenias-post-revolutionary-political-consensus.
Khachatryan N (2020) Protests and tensions escalate around Amulsar Gold Project in Armenia. https://www.bellingcat.com/news/2020/08/15/protests-and-tensions-escalate-around-amulsar-gold-project-in-armenia/
Martirosyan S, Khachatryan T (2020) Armenia's worsening air quality: particulate pollution in Yerevan Tops Moscow. https://hetq.am/en/article/112482
Mejlumyan A (2019a) The government recently began to crack down, and loggers–who say it's their only way to make a living–blocked roads to protest. https://eurasianet.org/armenian-police-protesters-clash-over-illegal-logging
Mejlumyan A (2019b) Pashinyan gives go-ahead to controversial Armenian mine project. https://eurasianet.org/pashinyan-gives-go-ahead-to-controversial-armenian-mine-project
Mejlumyan A (2022) Armenia signals readiness to restart controversial mine project. https://eurasianet.org/armenia-signals-readiness-to-restart-controversial-mine-project
Muradyan V (2021) Armenia takes on new waste management practices. https://evnreport.com/raw-unfiltered/armenia-takes-on-new-waste-management-practices/
Nieczuja-Ostrowski P (2016) Państwo i społeczeństwo wobec wyzwań i przemian współczesności. In: Nieczuja-Ostrowski P (ed) Armenia. Dziedzictwo a współczesne kierunki przemian kulturowo-cywilizacyjnych. Fundacja na rzecz Czystej Energii, Poznań, pp 696
Ordyan G, Pomieciński A (2021) Wielki zryw małego narodu. Zapomniany ruch karabaski (1988–1990). Roczniki Humanistyczne LXIX 2:193–212
Pomieciński A (2022) The Armenian genocide. Extermination, memory, sacralization. In: Bogumił Z, Yurchuk Y (eds) Memory and religion from a postsecular perspective. Routledge, London, New York, pp 69–88
Pomieciński A, Tadevosyan A, Fedorowicz K, Ordyan G (2022) Ludność cywilna wobec wojny w Górskim Karabachu. Antropologia straty i cierpienia. Instytut Badań nad Polityką Europejską, Gdańsk
Raport Armenia Bankwatch Network (2020) Amulsar: human rights violationsand environmental Negligencein the search for gold. https://www.fidh.org/IMG/pdf/armania_report_en.pdf
Raport EcoLur Press Club (2019) German journalists presented environmental problems in Armenia. https://www.ecolur.org/en/news/mining/german-journalists-presented-environmental-problems-in-armenia/11773/
Raport Trading Economics (2022) Armenia. Stopa bezrobocia. https://pl.tradingeconomics.com/armenia/unemployment-rate
Sanasaryan H (2018) AMULSAR: the criminal indulgence of the relevant bodies. http://bnaban.am/amulsar-hansavore/
Sargsyan L (2019) Mining sector's problems in Armenia: case of Amulsar. https://www.researchgate.net/publication/357811785_Mining_sector's_problems_in_Armenia_case_of_Amulsar
Saroyan G (2016) Depoliticising protests in Armenia. https://www.opendemocracy.net/en/odr/depoliticising-protest-in-armenia/
Vardanyan A (2020) Principal ecological threats of the mining industry inArmenia and its political and legal background. Succow Foundation, Greifswald. https://www.succow-stiftung.de/fileadmin/Ablage/Projekte/Forschung_Weiterbildung/Anna_Vardanyan_paper_MD.pdf
Veliyev J, Manukyan S, Gvasalia T (2020) Perspectives on peace in the South Caucasus through the lens of environmental activism. https://caucasusedition.net/perspectives-on-peace-in-the-south-caucasus-through-the-lens-of-environmental-activism/
Zdravomyslova E, Voronkov V (2002) The informal public in Soviet society: doublemorality at work. Soc Res 69(1):97–117
http://www.armecofront.net/lrahos/inchu-chi-kareli-shahagorcel-amulsary/
https://www.europarl.europa.eu/meetdocs/2009_2014/documents/envi/dv/envi20130925_info-cyanide_/envi20130925_info-cyanide_en.pdf
http://geopolitics.am/archives/15567

https://hetq.am/en/article/23886
https://www.lydianarmenia.am/eng/pages/About%20the%20mine/70/
https://www.lydianarmenia.am/eng/pages/eng/pages/amulsar-heap-leaching/124/
https://www.netinbag.com/pl/manufacturing/what-is-gold-cyanidation.html

Adam Pomieciński is Professor at the Institute Anthropology and Ethnology Adam Mickiewicz University in Poznań (Poland). His main research interests are Armenian studies, culture of ethnic and religious minorities in the Caucasus and problems of social movements. His publications include *Alter-Globalists. Anthropology of the Global Justice Movement (Alterglobaliści. Antropologia ruchu na rzecz globalnej sprawiedliwości)* (Adam Mickiewicz University Press 2013), the co-authored *National Minorities in Poland. An Anthropological Outline* (2019) and authored *Aliens in Their own Homeland? Syrian-Armenians in Armenia (Obcy we własnej ojczyźnie? Syryjscy Ormianie w Armenii)* (2017), *Syrian Armenians in Armenia. Adaptation-Reintegration-Diaspora* (2020), The Armenian Genocide: Extermination, Memory, Sacralization (Routletge 2022).

Chapter 5
On Cultural Direction of Socio-Ecological Transformations: Lessons from Degrowth and *Buen Vivir/sumak kawsay*

Buen Vivir/Sumak Kawsay

Katharina Richter

Abstract This chapter brings degrowth into conversation with *Buen Vivir/sumak kawsay*, an alternative to development from Ecuador. The Anthropocene is a crisis marked by multiple ecological crises, but also by dualistic and hierarchical structures of oppression. It's a civilisationary crisis that needs to be confronted in all its intersecting dimensions. Anthropocentrism is one of the defining features of this new geological epoch, and stands in the way of more profound socioecological transformations towards ecological sustainability and social wellbeing. This chapter therefore generates an inter-epistemic dialogue between *Buen Vivir/sumak kawsay*, an Andean-Amazonian conceptualisation of Good Living, and degrowth, a social movement from the Global North that advocates a democratic and redistributive reduction of affluency-based consumption and production patterns in line with social and ecological boundaries. The chapter is based on research carried out in Ecuador in 2020 into *Buen Vivir/sumak kawsay* in practice. The dialogue between these two projects serves to overcome remnants of anthropocentrism in degrowth thought and practice. Reciprocal practices with the non-human world, observed in Ecuador, can give impetus to the cultural direction of socioecological transformation processes, alongside socioeconomic reforms and policies.

Keywords Degrowth · Buen Vivir · Anthropocentrism · Anthropocene · Political ontology · Cultural politics

5.1 Introduction

Conversations and analyses about how to transform our economies and societies towards ecological sustainability and social wellbeing are characterised by concerns over modes of production, consumption levels, working hours, wealth distribution,

---

K. Richter (✉)
University of Bristol, Bristol BS8 1TL, United Kingdom
e-mail: K.Richter@bristol.ac.uk

M. K. De Santo and S. E. Domptail (eds.), *Degrowth Decolonization and Development*,
https://doi.org/10.1007/978-3-031-25945-6_5

the nature of money and debt, as well as economic practices, behaviours and organisational models. Yet, there is a sense that this picture doesn't represent the whole story, that we need more profound psychological and cultural changes alongside sweeping socioeconomic transformations. Dismantling, for instance, the dualistic juxtaposition of nature/culture is a fundamental condition for creating lasting and impactful socioecological transformations. This chapter brings together degrowth and *Buen Vivir* in an effort to shine light on the role of cultural politics in sustainability transformations.

Buen Vivir is based on indigenous Andean-Amazonian systems of knowledge and socio-economic organisation that centre reciprocal relationships in the human world, and between the human and non-human world. The translation of these principles into modern categories resulted in the enshrinement of the right to Good Living (*Buen Vivir* in Spanish) and Rights of Nature in the 2008 plurinational constitution of Ecuador. The post-development literature refers to it as an alternative *to* development (Acosta 2015). Development here is understood broadly as the post-1945 classification of the world into developed and developing countries, and its associated hegemonic bureaucratic and administrative structures, such as the World Bank or the United Nations Development Programme. As such, *Buen Vivir* attempts to undo the totalising and hegemonic logic of the modern development project based on economic growth, trade liberalisation and resource extraction. Instead, in its grassroots, decolonial sense, it proposes establishing and consolidating a Community Solidarity Economy, Rights of Nature, and interculturality and plurinationality.[1]

Degrowth, by comparison, originated in Western Europe in the early 2000s, although its core ideas have since spread to affluent countries and communities across the world. Degrowth is a social movement, and combines thought and action into an 'activist science' (Martinez-Alier et al. 2014). Its initial, methodologically individualistic concern lay with 'frugal abundance' and 'voluntary simplicity' as a solution to overconsumption and productivism (Akbulut 2021). Environmental justice movements from the Global South, as well as feminists, post/decolonial thinkers, and anti-racist activists, pushed degrowth to also focus on care, common resource governance, and the valuation of diverse knowledges (Paulson 2021; Barca et al. 2019, 3–4; Nirmal and Rocheleau 2019). Nevertheless, degrowth and *Buen Vivir* are based on different epistemological approaches. A dialogue between the two can make a useful contribution to addressing the intersecting challenges of the Anthropocene—the new geological epoch marked by human activity and structures of domination (Lander 2019, 15; Zalasiewicz et al. 2011). What can degrowth gain from the weaving of such an 'ecology of knowledges'?[2]

---

[1] Interculturality refers to the political and ideological principle of mutual respect for, and equitable relations between, the different cultural practices of Ecuador's diverse ethnic groups and nationalities (CONAIE 2012, 55). Plurinationality refers to the exercise of collective rights and self-determination by the Ecuadorian indigenous nationalities within the modern nation state.

[2] There are, of course, lessons to be learned from degrowth for the practice and theory of BV/sk, for example, in relation to its political economy analysis and operationalisation of reciprocity. Analysing these lessons, however, would go beyond the scope of this chapter.

An 'ecology of knowledges' brings together, and bridges some of the gaps between, different knowledge systems in order to overcome their respective limitations in addressing the intersecting challenges of the Anthropocene (Santos 2007). For degrowth, such an inter-epistemic dialogue with *Buen Vivir* is pertinent because while *Buen Vivir* centres the importance of cultural practices, symbolisms and rituals, there are fewer cultural analyses of degrowth processes than socioeconomic studies. The latter convincingly model quantitative and qualitative changes towards social equity, wellbeing, and ecological sustainability. Combined, these analyses produce pervasive arguments and visions for a transformation towards a low carbon, post-growth future. However, because of an under-prioritisation of cultural analyses, some parts of degrowth scholarship have remained grounded within a Eurocentric focus.

This Eurocentrism is characterised, amongst others, by anthropocentric conceptualisations of nature, and Western notions of time based on labour market logics, rather than agricultural or cyclical notions (Paulson 2014, 46; Nirmal and Rocheleau 2019). These perspectives stand alongside cultural studies of degrowth in relation to Ubuntu (Ramose 2014), the *khat* economy in Madagascar (Gezon 2017) or social property claims to springs and trees in Brazil (DeVore 2017). Efforts to decolonise degrowth and diversify its geographical and epistemological reach are therefore well under way (Deschner and Hurst 2018; Varvarousis 2019). This chapter complements these efforts by confronting remnants of degrowth anthropocentrism by making connections with *Buen Vivir* in Ecuador.

In Gramscian terms, cultural analyses can point us to ways in which common senses, that is, the dominant and often unconscious ways in which we perceive the world, can be mobilised, for instance, to challenge the dominant view of nature as a resource (Hoare and Smith, 1971, p. 322 in D'Alisa 2019, 250). Strengthening cultural analyses is important for degrowth because potential allies from the Global South, for instance, have rejected degrowth allyship because of its Eurocentrism and anthropocentrism (Nirmal and Rocheleau 2019). Decolonising degrowth would therefore expand degrowth's engagement with the multiple, intersecting dimensions of the Anthropocene, such as anthropocentrism and hierarchical knowledge production. Not engaging with anthropocentrism amounts to a failure to engage with a defining feature of the Anthropocene. The effort to decolonise degrowth, that is, make visible and challenge ongoing colonial logics of domination is therefore not just a moral endeavour, but an attempt to consolidate degrowth thinking with regards to the Anthropocene's major intersecting dimensions. Consequently, I approach *Buen Vivir/sumak kawsay* (BV/sk), a Kichwa alternative to development from Ecuador, from the vantage point of degrowth thinking, and think about the practice and theory of degrowth from the perspective of BV/sk. To illustrate these arguments, and show ways to decolonise degrowth that are inspired by the practice of BV/sk, the chapter will be structured as follows.

First, I will give a brief methodological overview. Second, I will analyse the prominence of anthropocentrism in degrowth thinking. Third, I will introduce *Buen Vivir*, and sketch the theoretical background and practical implication of the political ontology of BV/sk, on which subsequent arguments rest. Finally, the chapter concludes with some final remarks on the inter-epistemic dialogue between degrowth and *Buen Vivir.* The proceeding part will discuss the ways in which this research conceptualises the Anthropocene and what this means for decolonising degrowth.

5.2 Research in Times of Civilisationary Crisis

This chapter gives analytical weight to forms of political and socio-economic organisation that centre socio-ecological wellbeing, but which have been marginalised from (academic) knowledge production. To work against this marginalisation on an epistemological and theoretical level, this chapter conceptualises the Anthropocene—the current geological epoch defined by human impact on the planet—as a *civilisationary* crisis. On the one hand, this notion by Edgardo Lander (2019, 15) highlights the progressively escalating devastation of the conditions that create and reproduce life. On the other hand, it recognises that the Anthropocene is not just defined by ecological crises, but by the multiple, intersecting civilisationary patterns of sexism, racism, anthropocentrism, classism etc., characteristic of modernity.[3] Crucially, Lander also contends that science and technology, today's hegemonic approaches to knowledge, exacerbate, rather than solve, this civilisationary crisis (Lander 2019, 14). Green growth, speculative, large-scale negative emissions technologies or geoengineering are all symptoms of a tendency to prioritise technological approaches over socio-political solutions (Hickel and Kallis 2019). Thinking about today's intersecting crises as a *civilisationary* crisis has the potential to challenge those modern hierarchical, binary, and dichotomous analytical categories and knowledge structures that sustain unequal relationships between knowledge systems, between people, and between the human and non-human world. In terms of decolonising degrowth, examining degrowth through the lens of the civilisationary crisis facilitates locating and challenging any hierarchical and exclusionary analytical categories.

In early 2020, I interviewed 15 social leaders, community members and politicians, and observed and participated in municipal, indigenous, and NGO assemblies, as well as indigenous rituals and celebrations and public meetings. The locations were spread across Ecuador. Most of the interviews and public meetings took place in Quito, but I was also part of conversations and rituals with people in the Southern highlands of the country, for example, in Saraguro, Loja and El Tambo, Cañar province. I was also participant observer and conducted interviews at the 3rd Women's Assembly of the Ecuadorian Amazon, hosted by the Confederation of Indigenous Nationalities of the Ecuadorian Amazon (CONFENAIE) in Puyo, one of the Amazonian cities in the country's East. In Intag, I took part in the 25th anniversary celebrations of DECOIN, *Defensa y Conservación Ecológica de Intag*, an

[3] Defined here as a specific, that is, Euro-Atlantic, civilisationary model whose socio-cultural, psychological and economic structures, such as capitalism, heteronormativity, the nuclear family, subjectivities like the citizen, *homo economicus* or the consumer, a racialised and gendered division of salaried and non-salaried labour exploitation, and the nature/culture binary, have their origins in Renaissance Europe and its constitutive colonial conquests and exploitation in the tropics and elsewhere (Escobar 2010; Lugones et al. 2008; Quijano 2000).

environmental NGO in the province of Imbabura, northeast of Quito, and observed municipal assemblies and roundtables on environmental issues. Finally, I participated in two fieldtrips on alternatives to development in Cayambe, located in the Northern province of Pichincha, and Nabón, Azuay province, led by academics of the Simon Bolivar Andean University in Quito and the University of Cuenca in Southern Ecuador respectively. The data collection process focussed on the lived experiences and practices of the rural (and urban) participants of my research, including their reciprocal practices, rituals, and cultural and agricultural practices.

The interview questions and strategy critically reflected on interviewing in a context of power imbalance, since colonialism and coloniality historically link research with the appropriation of indigenous knowledge (Smith 2012). Though this research largely avoids producing "discourse under conditions of unequal power" (Escobar 1995, 9) that would essentialise and homogenise interviewees (Ibid; Mohanty 1988), knowledge was nevertheless generated under conditions of asymmetric power relations, reflected, for instance, in my own positionality as a white researcher and privileged degrowth activist-scholar from a Higher Education institution in the Global North. To engage with these power imbalances, and to avoid appropriating indigenous knowledge, my methodology included engaging with indigenous scholarship prior to, during and after fieldwork. Amongst others, my research methods produced novel insights into the political ontology of BV/sk, which have informed the subsequent arguments in this chapter. The following part will briefly elaborate on degrowth's anthropocentrism, before introducing BV/sk in more detail.

5.2.1 Colonial Natures

How does anthropocentrism in degrowth thought manifest itself? Degrowth articulates ecological, political, and cultural critiques towards the primacy of economic growth. This chapter focuses on the part of degrowth scholarship that articulates an ecological critique of growth, based on ecological economics and thermodynamics. Inspired by ecological economics, the economy is argued to be embedded in global ecosystems which reproduce life on Earth. Societies, as argued in one of the founding texts in ecological economics, "receive inputs from the earth, the atmosphere, and the waters, and [...] give outputs into these reservoirs" (Boulding 1966, 2). To conceptualise nature as resources (in- and outputs) abstracts and objectifies sentient beings and abiotic elements, which ultimately detaches them of cultural meaning, local values, and the colonial histories that have carved and moulded entire landscapes and societies. The part of degrowth scholarship that is concerned with economic modelling of the macro- and microeconomics of postgrowth transformations in relation to money, work and wellbeing, utilises these abstractions (for examples of degrowth analyses that construct nature as exclusively non-human ecosystems, see Akizu-Gardoki et al. 2020; Andreoni and Galmarini 2014; D'Alisa and Cattaneo 2013; Heikkinen 2020; 2018; Jackson and Victor 2020; O'Neill 2012; Victor 2012; Videira et al. 2014; for a critique of the nature/cultural dualism in ecological economics, see Kolinjivadi 2019). Though the language since the 1960s has moved on to that of 'materials', rather than 'resources', there are two wider points to make in relation to this conceptualisation of nature. One, it is arguably sustained by a mechanistic, materialist view on nature; and two, it universalises a particular, scientific, and European-Atlantic view on nature—a tendency expressed by Latin American thinkers in terms of the 'coloniality of nature' (Achinte and Rosero 2016; Alimonda 2011; Machado 2010). The following paragraphs will substantiate these two assertions.

The first point follows Carolyn Merchant (2020) in arguing that historically, the drastic and violent economic changes that accompanied the emergence of merchant capitalism in the fifteenth and sixteenth centuries heralded fundamental changes in European cosmology. These economic changes were accompanied, and in part enabled by scientific developments prior to and during the Industrial Revolution. The privatisation of forests and fens that had previously been managed communally was followed by technological advancements in metallurgy, mining and steam power in the eighteenth and nineteenth centuries (Merchant 2020, 37). Industrialists gained control over, and centralised 'resources' (inputs, followed by outputs) that had to be dug out of the Earth, felled, smelted or burned (Malm 2015). Nature, once regarded as a living, female, and nurturing force, was subdued, dominated, and finally seen as dead, inert matter.

Once nature had become an inanimate machine, its individual parts could be predicted, controlled, and manipulated by scientific intervention (Allan 2018, 93; Merchant 2020, 37, 41, 74, 80, 156, 186, 263). The death of nature was a fundamental change in the way people in Europe thought about their place in the universe. They came to believe that the human condition can be improved by acquiring knowledge about a world made of 'materials' and the mechanistic forces governing their use and extraction (Allan 2018, 11, 95, 165). The fundamental question of what exists, and how it exists, is what the term 'ontology' refers to (Allan 2018, 11). Politically, these ideas gained traction via public intellectuals such as Francis Bacon or William Temple, and laid the foundation for twentieth century notions of progress and economic growth. The mechanistic understanding of nature sanctioned resource exploitation and motivated Western colonial conquests.

Similarly, modernity/coloniality scholars argue that the civilisationary crisis is rooted in the imperial discoveries of 'Nature', the Orient(al) and the 'Savage' as lucrative assets and economic resources (Lander 2018, 48–50; Santos 2006, 125). Their careful control and domination transformed "wild"peoples and nature into labourers and plantation economies (Pluymers 2011). Colonial domination ultimately made these societies and landscapes suitable for civil, that is, English occupation. Plantations and factories allowed for easy exchange—and by extension, commodification—of slaves, sugar cane, machine parts, and workers (Mintz, 1986 in Tsing 2015, p. 38). Nature became the inherent opposite of culture and cultured (Anglo-Saxon) society. The separation of nature from culture is a core, binary ontological assumption within the modern, scientific knowledge system (Mignolo and Carballo 2014, 30). The colonisation of nature and natural spaces by economic interest, industry and extraction is one aspect of the 'coloniality of nature'. However, there is a second, far wider-reaching epistemological meaning of the term.

In addition to nature's subordination to market interests, the notion of 'coloniality of nature' also refers to the domination exercised by the discourse of nature as separate from culture over non-dualist ways of relating to the non-human world (Alimonda 2011, 47–50). This hegemonic discourse on the environment determines who may know and exploit, and who may protect and defend nature (Achinte and Rosero 2016). It marginalises struggles such as those in the Colombian Pacific, where social movements redefine modern-scientific notions such as 'biodiversity' to encompass *territory plus culture,* set within specific power relations (Escobar 2000, 131–32; Porto-Gonçalves 2009). It thereby forecloses biophysical and sociocultural ways of *producing* territory—and concomitant knowledge systems—that regard territory as something akin to nature-cultures, that is, political spaces that are both nature *and* culture (Paulson 2014, 46). Over the course of time, the geopolitics of knowledge production has standardised and universalised technocratic knowledge production as scientific.

In the twenty-first century, the combination of technoscientific knowledge production with the discursive power of neoliberalism marginalises indigenous knowledges in the fight against climate change and produces a climate eco-governmentality based on REDD+, Payments for Ecosystem Services, and carbon markets (Ulloa 2012, 18). The economic models and policy proposals of degrowth are far removed from the neoliberal governmentality of mainstream climate change mitigation and adaptation. However, the underlying, mechanistic and materialist ontology that mandates the abstraction of nature into economic models can also be found in the ecological economics tradition of degrowth. As such, this section of degrowth scholarship unwittingly perpetuates the coloniality of nature by universalising an idea of nature that has its origins in early and classical modern (pre-twentieth century) Europe (Richter 2019). Decolonial work requires acknowledgment that the idea of nature as separate from culture is a Euro-Atlantic construct (Mignolo and Carballo 2014, 133). Latin American activists, thinkers, and social movements have gone further and put forward a wide array of proposals on how to decolonise the construction of nature as separate from culture. One such proposal, and political project, is *Buen Vivir/sumak kawsay* (BV/sk). The next part will analyse BV/sk in practice, before concluding this chapter with an inter-epistemic dialogue between BV/sk and degrowth that attempts to dislocate the anthropocentric aspects of degrowth thought and practice.

5.3 A Brief History of Buen Vivir/Sumak Kawsay in Ecuador

The constitutional right to *Buen Vivir* (Good Living) in Ecuador is based on the Kichwa concepts of *sumac kowsay*, or *alli kawsay*, which translate into 'life in excellence or plenitude'. The use of these concepts as a pedagogical principle by indigenous movements in the fight against illiteracy, poverty and discrimination is documented in texts dating back to the beginning of the twentieth century (Inuca 2017). The 2004 founding of theIntercultural "Pluriversity" as a self-governed institution for higher education for the Indigenous Nationalities, Amawtay Wasi, based on the principles of *sumak kawsay*, is a continuation of that use. More broadly, *Buen Vivir* has become an umbrella term for a confluence of practices and principles of historically racialised and marginalised communities working towards equilibria with the self, community and nature (Astudillo 2020, 247). The following paragraphs will discuss the details of these proposals, as well as the key political trajectory of BV/sk in Ecuador from the 1980s onwards.

During the early 1980s, the indigenous organisations of the Ecuadorian highlands consolidated the political and ideological meaning of *alli kawsay* in the fight against oppression and exploitation. This fight was based on the modes of living and socio-political organisation of the Kichwa nationalities, based on close connections to the land, within families, and the spiritual realm (Inuca 2017). The struggle for "land, life in plenitude and liberty ("*con unidad por la tierra y la vida hermosa hasta liberarnos*") culminated in the creation of a system of political and social organisation based on plurinationality and interculturality—two central components of *sumak kawsay* as a political project (Ibid). In the Ecuadorian Amazon, the notion of post-development was brought to indigenous nationalities by anthropologists and development institutions (Hidalgo-Capitán and Cubillo-Guevara 2017). During the 1990s, *sumak kawsay* subsequently emerged as a counterproposal to the World Bank's sustainable development paradigm and was disseminated to Andean intellectual and social movement circles (Ibid). Textual evidence and my research demonstrates that this was not an invention of tradition, but a recuperation of praxis and memory, as one of my interviewees describes:

In our communities, we do still have *sumak kawsay* because we are connected with nature; because we can still talk with each other in confidence [over the fire, as a ritual], sharing our experiences; because we can share our pain and share our food. It's all there. Because we still have our celebrations, all of it.

BV/sk as a political proposal entered the public sphere in the early 2000s. This coincided with a recognition of diversity as a useful political strategy by the Ecuadorian left-wing movements, whose struggles subsequently converged with those of indigenous and afro-descendant's movements (Amawtay Wasi 2004; Cortez 2011). These developments were part of wider political trends across Latin America under the so-called "pink-tide" of centre-left governments. In Ecuador, PAIS Alliance under Rafael Correa convened a constituent assembly to draft a new constitution based on greater economic state control over key resources. These processes culminated in the constitutionalisation of *sumak kawsay* as *Buen Vivir*, or the right to Good Living in the country's new constitution, approved via referendum in 2008. Until the mid-2000s, *Buen Vivir* was incorporated as a 'human development' principle into Correa's development plans, which followed Sen's capabilities approach. The planning documents inscribed *Buen Vivir* as the overarching development goal based on economic diversification, food sovereignty and the economy of care (Senplades 2007, 2009). However, indigenous territorial and plurinational rights came to be subordinated to Correa's neo-extractivist development and modernisation model, pushed through under the guise of *Buen Vivir*.

Between 2007 and 2017, the Correa government systematically weakened indigenous, environmental and other social movements as political actors by oppressing dissent towards its extraction-based growth-model (Novo 2013). Examples include the closure of the Pluriversity of Amawtay Wasi in 2014, weakening of bilingual education, and granting mining concessions for oil and gas exploitation in indigenous territories. The meaning of *Buen Vivir* is now bifurcated: on the one hand, '*Buen Vivir*' became a government discourse and tool for domination, while it remained a grassroots, decolonial project in the form of '*sumak kawsay*' (Gudynas 2014, 29).[4] The latter continues to be a political platform that unites the Ecuadorian indigenous movements in the struggle towards self-governance, equality, and wealth redistribution.[5]

5.4 The Realities of Reciprocity

The inter-epistemic dialogue between degrowth and BV/sk draws on the field of political ontology. Political ontology coalesces arguments made within political ecology, anthropology, post-humanism, science and technology studies and indigenous studies into the assertion that struggles over land and modes of living also involve struggles over reality. Ontological struggles particularly pertain to indigenous communities, organisations and movements. From this perspective, the plurinational and Rights of Nature elements of the 2008 Ecuadorian constitution successfully overturned the previous, neoliberal constitution that was based on a Cartesian worldview and a homogenous, mono-ethnic citizenship model. By extending (political) agency to nature, BV/sk therefore generates an 'ontological conflict' (Blaser 2013) that confronts the mechanistic view of nature as a resource and separate from society. BV/sk thereby also displays an "ontological disobedience" (Burman 2017a, 168) towards the underlying economic interests behind this dualistic worldview that mandate mineral, oil and gas concessions. This is not a mere abstract assertion. Ontology, that is, what we believe to be real, is made visible through world-making practices, that is, the every-day practices that are constitutive of an ontology (Strathern 2019). I set out to observe these practices in Ecuador, which I argue create a relational ontology that connects human wellbeing to natural and spiritual wellbeing. Relational ontologies compose those worlds which do not rely on a divide between nature and culture but instead are produced by the interactions

[4] This chapter uses the abbreviation BV/sk to acknowledge the fact that the term *Buen Vivir* has evolved beyond its original Kichwa meaning without erasing its indigenous provenance. As such, 'BV/sk' recognises this political project as a *civilisationary alternative* (Richter 2022, 277).

[5] Use of *Buen Vivir* in a non-government sense, however, isn't restricted to the indigenous movements – the cooperative housing project *Alianza Solidaria* in Quito is known as the "*barrio del Buen Vivir*" (neighbourhood of Good Living) and is built on collective labour, ecological restoration and community organisation (Richter 2022, 320–22).

between human and non-human actors (Escobar 2019; for studies in Latin American relational ontologies, see Blaser 2013; de la Cadena 2015).

This school of thought does not contest what is real, but rather aims to highlight the ontological dimension of global coloniality, which is to say, the conflict between different realities in a world that in the tradition of the Zapatistas contains many worlds—a *pluriverse* (EZLN 2005; Burman 2017a, 160–97). One of my participants from the Amazonian Sarayaku nation illustrates this argument:

We, the people of Sarayaku, live in two realities. The Western one and the traditional. The community, the traditional, is our form of life. The system has obliged us to know the reality of the Western world. For example, in speaking, Spanish, Castilian, the way we speak, this isn't our own language. We have the Kichwa language. This obliges us to communicate. To make them listen to us, because it's the only way to express ourselves.

Political ontology asks whose realities count and are allowed to be 'real' (Burman 2017b). In that sense, a rainforest can be a cosmos, a community of living beings which includes the invisible spirits of the forest, as well as a resource pool for lumber, copper, crude oil, gold or commodifiable genetic diversity. The hierarchical nature/culture binary is ultimately a cultural belief too. Its world-making practices, however, are causing dangerous environmental change on a planetary scale. The value chains of the global political economy turn this binary into a reality for most of humankind, regardless of whether their own world-making practices are constitutive of this belief. Nature can both be conceived of as a resource, *and* as ancestral spirits and earth-beings. However, the struggles over being, knowledge and reality aren't theoretical—they are composed of material and political conflicts over territorial integrity and autonomy, forms of decision-making and politics, struggles against racism, sexism, ecocide etc. (Burman 2017a, 162). Consequently, *sumak kawsay* in practice can be found in efforts to recuperate ancestral farming practices and seeds, seasonal indigenous rituals and celebrations, and the traditions, clothing, and self-governance of the Ecuadorian indigenous nationalities. The following are a few examples of world-making practices, in other words, socioeconomic behaviours and ritualistic patterns that I observed and which create a relational ontology.

I noticed that when I was talking to others over a fire, I was occasionally offered to rub my hands with an ointment made out of herbs and other plants, that were grown, as

I was told, in the speaker's *chakra* using agroecological farming methods.[6] In recalling some of its crops and their provenance, this small act had the effect of reminding those around the fire of their intimate connection with nature. I also observed the ritualistic spilling of *chicha* or other alcoholic beverages on the ground as an offering to *Pachamama*, or Mother Earth. I argue that in addition to holding *Pachamama* in ongoing awareness, this act also produces care and affection for her in rural, agricultural-dependent communities. Translating care and affection for *Pachamama* into agroecological practices, however, isn't always the most socially

[6] A *chakra* is an allotment, and the spiritual and productive centre of communal and family life (FAO 2020).

and economically advantageous choice. The Agroecological Women's Association of Sumak Mikuna ("Excellent Food") in El Tambo, Cañar province, fought against prejudice towards organic food production when they set up the association in response to being boycotted at local markets. In addition, their smaller, more costly fruits and vegetables are less competitive compared to produce grown with agrochemicals. Nevertheless, the associates relate their children's and their own health to the quality of the topsoil under their care. Quinoa and amaranth bars are sold at local schools, and the association's sales and mutual exchange network stretch across the southern provinces of Ecuador up to the coast. The political ontology of BV/sk can therefore be said to be founded on relationality.

Finally, the agroecological practices of another producer, this time in Cayambe, Pichincha, again illustrate the practice of maintaining reciprocal relationships with the non-human world. Rather than growing only vegetables, she intersperses her *chakra* with specific species of trees to attract—and nourish—specific species of birds. By fertilising the vegetables growing underneath the trees, and providing rain and hail cover, I would argue that this practice improves agricultural yield by natural means. However, for the producer it also is a form of sharing part of her agricultural space with non-human species so that these may fulfil their needs, which are considered legitimate needs, rather than "theft" of crops destined for human consumption. The political ontology, and constitutive world-making practices such as these embody reciprocal relationships with non-human communities.

5.5 An Inter-Epistemic Dialogue Between Degrowth and Buen Vivir/Sumak Kawsay

This section will lay out why an engagement with the world-making practices of BV/sk is a fruitful endeavour for degrowth, and may serve to challenge latent anthropocentrism in its ecological growth critique. The socioeconomic and ritualistic behaviours described in this chapter maintain ecological equilibria, which limit harmful growth. Balancing relationships between the human and non-human world to maintain these equilibria satisfies basic needs and generates material and spiritual wellbeing in people (Churuchumbi 2014). *Pachamama* thereby isn't a mere symbol of fertility, but a living being reminiscent of the image of a female, nurturing Earth in pre-modern, pre-capitalist European cosmologies (Merchant 2020, Chap. 1). Both *Pachamama* and the image of a female Earth in pre-modern Europe act as a normative restraint to resource extraction and environmental degradation. This is not a historicist argument—the protagonists of BV/sk are to varying degrees connected to the modern state and capitalist modes of production (Strathern, 2004 in de la Cadena 2015, 32). It might be argued that concerns over limits might seem difficult to inscribe into communities whose lives are centred around relational, rather than linear and fundamentally binary cosmologies (Inuca 2017). Yet, in Cayambe, for example, export-oriented flower production threatens groundwater level and quality.

The flower industry moreover offers only precarious and insecure working conditions. As such, there is very much an understanding that these industrial activities have harmful socio-ecological effects that ought to be resisted and limited. Furthermore, agroecological production doesn't take place in a social vacuum—its (women) farmers are affected by interlinking patriarchies, the drain of outward migration and state-sanctioned mineral, oil and gas extraction. But rather than perceiving the world as constrained and scarcity-driven, the protagonists of BV/sk see the world as abundant—as long as everybody makes space for others to exist. Relational worldviews and their accompanying world-making practices embed the human into the non-human world. They are antithetical to the worldview that perceives humankind as the dominators and masters of an inert, controllable nature.

To conclude, relational worldviews based on connectedness with nature produce reciprocal relations with the non-human world. Contrasting them with the anthropocentric, dualistic conception of the environment as materials points to the enormity of the task ahead. If our underlying worldviews have the ability to produce ecological equilibria that constrain harmful productivism, we need to have more profound debates on the role of culture in socioecological transformations. From this vantage point, to change the way we think, act and feel ("*sentipensar*") towards the natural world is a crucial task in the transformation towards a low-carbon, socially just future.

5.5.1 A Pluriverse Avenue for the Cultural Politics of Degrowth

Degrowth thinkers have presented alternatives to mechanistic and anthropocentric views on nature from the Global South (e.g. Gezon 2017). Amongst others, these demonstrate that territories are biophysical and political entities that cannot be split into just 'culture' or 'nature'. This chapter has contributed to these efforts, and shown that reciprocal practices that I observed under the umbrella term of *Buen Vivir/sumak kawsay* in Ecuador made visible an underlying relational ontology that doesn't split nature from culture. These practices included offerings to Pachamama, the ritualistic use of herbal ointments, and multi-species care in agroecological production. Well-being is framed in both material and spiritual terms, and contingent on reciprocal relationships within the family, community, and with the non-human world. This relational view offers a pluriversalistic avenue for the cultural politics of degrowth.

The cultural transformation that needs to take place alongside socioeconomic changes will require us to think of ourselves as part of the living world, and perhaps even start practicing context-dependent and locally relevant acts of reciprocity with it. For the cultural politics of degrowth, it would be worth exploring the role of rituals and reciprocity with the non-human world in future discussions and research. There are three specific benefits to paying attention to the cultural politics of reciprocity with the non-human world. It would help overcome the remnants of anthropocentrism within degrowth, increase its attractiveness as an ally for global environmental justice movements, and finally, decolonise its aspects that remain grounded in a mechanistic view of nature. Generating solidarity not just between humans, but between the human and non-human world is a key task for changing the way we think of the living world, and eventually, act towards it. How to do so remains a topic for further research. To learn from BV/sk, degrowth thinkers and practitioners need to continue to be assertive about the role of culture and cultural change in creating impactful socioecological transformations.

Acknowledgements This chapter is based on conference presentations at two international degrowth conferences in 2021. It draws on PhD research conducted at the Goldsmiths Department of Politics and International Relations, 2016–2021. The research was supported by funding by the Royal Economic Society, the Goldsmiths Graduate School, and the British Federation of Women Graduates, for which the author is immensely grateful. The author would also like to thank Miriam Lang and Stephanie Eileen Domptail for their insightful comments on an earlier draft of this chapter.

References

Achinte AA, Rosero JR (2016) Colonialidad de La Naturaleza: ¿imposición Tecnológica y Usurpación Epistémica? Interculturalidad, Desarrollo y Re-Existencia. Nómadas 45:27–41

Acosta A (2015) El Buen Vivir como alternativa al desarrollo. Algunas reflexiones económicas y no tan económicas. Política y Sociedad 52(2):299–330. https://doi.org/10.5209/rev_POSO.2015.v52.n2.45203

Akbulut B (2021) Degrowth. Rethink Marx 33(1):98–110. https://doi.org/10.1080/08935696.2020.1847014

Akizu-Gardoki O, Kunze C, Coxeter A, Bueno G, Wiedmann T, Lopez-Guede JM (2020) Discovery of a possible well-being turning point within energy footprint accounts which may support the degrowth theory. Energy Sustain Dev 59(December):22–32. https://doi.org/10.1016/j.esd.2020.09.001

Alimonda H (2011) La colonialidad de la naturaleza: Una aproximación a la Ecología Política latinoamericana. In: Alimonda H (eds) La naturaleza colonizada: ecología política y minería en América Latina. Grupos de trabajo de CLACSO. Ediciones CICCUS: CLACSO, Buenos Aires, pp 21–55

Allan BB (2018) Scientific cosmology and international orders. Cambridge Studies in International Relations. Cambridge University Press

Amawtay W, Universidad Intercultural (2004) Aprender en la sabiduría y el buen vivir/Sumak Yachaypi Alli Kawsaypipash Yachakuna. Amawtay Wasi Pakta kawsaypak Sumak Yachay Wasi, Quito

Andreoni V, Galmarini S (2014) How to increase well-being in a context of degrowth. Futures 55(January):78–89. https://doi.org/10.1016/j.futures.2013.10.021

Astudillo JE (2020) Prácticas del Buen Vivir: experiencias en comunidades shuar, kichwa y manteña. Universidad de Cuenca; Ediciones Abya-Yala, Cuenca; Quito

Arturo E (1995) Encountering development: the making and unmaking of the third world. Princeton Studies in Culture/Power/History. Princeton University Press, Princeton, N.J

Arturo E (2019) Thinking-feeling with the earth: territorial struggles and the ontological dimension of the epistemologies of the South'. In: Knowledges born in the struggle. Routledge

Arturo E (2000) 'El Lugar de La Naturaleza y La Naturaleza Del Lugar: ¿globalización o Post-desarrollo?' In: Lander E (ed) La Colonialidad Del Saber: Eurocentrismo y Ciencias Sociales:

Perspectivas Latinoamericanas. Buenos Aires: Consejo Latinoamericano de Ciencias Sociales-CLACSO; UNESCO, Unidad Regional de Ciencias Sociales y Humanas para Améric Latinay el Caribe, pp 155–200

Barca S, Chertkovskaya E, Paulsson A (2019) The end of political economy as we knew it? from growth realism to Nomadic Utopianism. In: Chertkovskaya E, Paulsson A, Barca S (eds) Towards a political economy of degrowth. Rowman & Littlefield Publishers, London, pp 1–17

Blaser M (2013) Ontological conflicts and the stories of peoples in spite of Europe: toward a conversation on political ontology. Curr Anthropol 54(5):547–568. https://doi.org/10.1086/672270

Boulding KE (1966) The economics of the coming spaceship earth. Environmental quality issues in a growing economy. https://books.google.com/books?hl=en&lr=&id=3PDqBgAAQBAJ&oi=fnd&pg=PT476&dq=%22something+like+a+frontier.+That+is,+there+was+always%22+%22of+the+spherical+earth+and+a+closed+sphere+of%22+%22cylinder,+essentially+a+plane+wrapped+around%22+%22grips+with+the+ultimate+consequences+of+the+transition%22+&ots=IdeCsjKwZe&sig=u0mou--9AmyijVVIlwATBkjH3dU

Burman A (2017a) La ontologia política del vivir bien. In: de Munter K, Michaux J, Pauwels G, Cajías M (eds) Ecología y reciprocidad: (con)vivir bien, desde contextos andinos. La Paz: Plural Editores; [Oruro]: CEPA, Centro de Ecología y Pueblos Andinos, pp 155–176

Burman A (2017b) The political ontology of climate change: moral meteorology, climate justice, and the coloniality of reality in the Bolivian Andes. J Political Ecol 24(1):921–930. https://doi.org/10.2458/v24i1.20974

Churuchumbi G (2014) Usos Cotidianos Del Término Sumak Kawsay En El Territorio Kayambi. Universidad Andina Simón Bolívar, M.A., Quito

Claire D, Hurst E (2018) Decolonisation and degrowth. Degrowth.Info (blog), 2 January 2018. https://www.degrowth.info/en/2018/02/decolonisation-and-degrowth/

CONAIE (2012) Proyecto Político Para La Construcción Del Estado Plurinacional: Propuesta Desde La Visión de La CONAIE—2012–2024. Confederación de Nacionalidades Indígenas del Ecuador, Quito

David C (2011) La Construcción Social Del "Buen Vivir" (Sumak Kawsay) En Ecuador: Genealogía Del Diseño y Gestión Política de La Vida. Aportes Andinos. Universidad Andina Simón Bolívar, Quito, Sede Ecuador; Programa Andino de Derechos Humanos, 28 (January). http://repositorio.uasb.edu.ec/handle/10644/2788

de la Cadena M (2015) Earth Beings: ecologies of practice across Andean Worlds. Duke University Press Books, Durham

D'Alisa G, Cattaneo C (2013) Household work and energy consumption: a degrowth perspective Catalonia's case study. J Clean Prod Degrowth Theory Practice 38(January):71–79. https://doi.org/10.1016/j.jclepro.2011.11.058

DeVore J (2017) Trees and springs as social property: a perspective on degrowth and redistributive democracy from a Brazilian Squatter Community. J Polit Ecol 24:644–666

Eduardo G (2014) Buen Vivir: Sobre Secuestros, Domesticaciones, Rescatas y Alternativas. In: Oviedo A (ed) Bifurcación Del Buen Vivir y El Sumak Kawsay. Ediciones SUMAK, Quito, pp 25–45

Escobar A (2010) Worlds and knowledges otherwise: the Latin American modernity/coloniality research program. In: Mignolo WD, Escobar A (eds) Globalization and the decolonial option. Routledge, London, pp 32–55

EZLN (2005) Sexta Declaración de la Selva Lacandona. Enlace Zapatista (blog). http://enlacezapatista.ezln.org.mx/sdsl-es/

FAO (2020) Andean Chakra System of the Cotacachi Kichwas Communities. http://www.fao.org/giahs/giahsaroundtheworld/proposed-sites/latin-america-and-the-caribbean/chakra-andina/detailed-information/en/

Giacomo D (2019) The state of degrowth. In: Chertkovskaya E, Paulsson A, Barca S (eds) Towards a political economy of degrowth, Illustrated edition. Rowman & Littlefield Publishers, London, pp 243–57

Gezon LL (2017) Beyond (Anti) Utilitarianism: Khat and alternatives to growth in Northern Madagascar. J Polit Ecol 24:582–594
Heikkinen T (2018) An equilibrium framework for the analysis of a degrowth society with asymmetric agents, sharing and basic income. Ecol Econ 148(June):43–53. https://doi.org/10.1016/j.ecolecon.2018.02.007
Heikkinen T (2020) A study of degrowth paths based on the von Neumann equilibrium model. J Clean Prod 251(April):119562. https://doi.org/10.1016/j.jclepro.2019.119562
Hickel J, Kallis G (2019) Is green growth possible?' New Polit Econ:1–18. https://doi.org/10.1080/13563467.2019.1598964
Hidalgo-Capitán AL, Cubillo-Guevara AP (2017) Deconstruction and genealogy of Latin American good living (Buen Vivir): the (Triune) good living and its diverse intellectual wellsprings. Int Develop Policy 9(9):23–50. https://doi.org/10.4000/poldev.2351
Inuca JB (2017) Genealogía de Alli Kawsay/Sumak Kawsay (Vida Buena/Vida Hermosa) de Las Organizaciones Kichwas Del Ecuador Desde Mediados Del Siglo XX. Lat Am Caribb Ethn Stud 12(2):155–176. https://doi.org/10.1080/17442222.2017.1325101
Jackson T, Victor PA (2020) The transition to a sustainable prosperity-a stock-flow-consistent ecological macroeconomic model for Canada. Ecol Econ 177(November):106787. https://doi.org/10.1016/j.ecolecon.2020.106787
Kolinjivadi V (2019) Avoiding dualisms in ecological economics: towards a dialectically-informed understanding of co-produced socionatures. Ecol Econ 163(September):32–41. https://doi.org/10.1016/j.ecolecon.2019.05.004
Edgardo L (2018) The Bolivarian experience: a struggle to transcend capitalism. In: Lang M, König CD, Regelmann AC (eds) Alternatives in a world of crisis. Global Working Group Beyond Development. Rosa Luxembourg Stiftung, Brussels Office, Brussels, pp 46–89
Lander E (2019) Crisis Civilizatoria: Experiencias de Los Gobiernos Progresistas y Debates En La Izquierda Latinoamericana. Guadalajara, Jalisco: Centro María Sibylla Merian de Estudios Iberoamericanos Avanzados en Humanidades y Ciencias Sociales (CALAS): Editorial Universidad de Guadalajara
Lugones M, Jiménez-Lucena I, Tlostanova M (2008) Género y Descolonialidad. In: Mignolo WD (ed) Ediciones del Signo and Globalization and Humanities Project (Duke University). El Desprendimiento. Buenos Aires: Del Signo
Machado H (2010) La "Naturaleza" Como Objeto Colonial. Una Mirada Desde La Condición Eco-Bio-Política Del Colonialismo Contemporáneo. Boletín Onteaiken 10(November):35–47
Malm A (2015) Fossil capital: the rise of steam-power and the roots of global warming. Verso Books, London, New York
Martinez-Alier J, Anguelovski I, Bond P, Del Bene D, Demaria F, Gerber JF, Greyl L et al (2014) Between activism and science: grassroots concepts for sustainability coined by Environmental Justice Organizations. J Politi Ecol 21(1). https://doi.org/10.2458/v21i1.21124
Merchant C (2020) The death of nature: women, ecology, and the scientific revolution. HarperCollins, Digital New York
Mignolo WD, Carballo F (2014) Una Concepción Descolonial Del Mundo: Conversaciones de Francisco Carballo Con Walter Mignolo. El Desprendimiento. Ediciones del Signo and Center for Global Studies and the Humanities, Duke University. https://www.edicionesdelsigno.com.ar/el-desprendimiento/una-concepcion-descolonial-del-mundo/
Mohanty CT (1988) Under western eyes: feminist scholarship and colonial discourses. Fem Rev 30:61–88. https://doi.org/10.2307/1395054
Nirmal P, Rocheleau D (2019) Decolonizing degrowth in the post-development convergence: questions, experiences, and proposals from two indigenous territories. Environ Plann E Nat Space 2(3):465–492. https://doi.org/10.1177/2514848618819478
Novo CM (2013) The backlash against indigenous rights in Ecuador's Citizen's revolution. In: Eisenstadt TA, Danielson MS, Corres MJB, Polo CS (eds) Latin America's multicultural movements: the struggle between communitarianism, autonomy, and human rights. Oxford University Press, Oxford, pp 111–132. https://doi.org/10.1093/acprof:oso/9780199936267.003.0033

O'Neill DW (2012) Measuring progress in the degrowth transition to a steady state economy. Ecol Econ Econ Degrowth 84(December):221–231. https://doi.org/10.1016/j.ecolecon.2011.05.020
Paulson S (2014) Political ecology. In: D'Alisa G, Demaria F, Kallis G (eds) Degrowth: a vocabulary for a new era. Routledge, Abingdon, Oxon, pp 45–48
Paulson S (2021) Decolonizing technology and political ecology futures. Polit Geogr 88(January):102369
Pluymers K (2011) Taming the Wilderness in sixteenth- and seventeenth-century ireland and Virginia. Environ Hist 16(4):610–632. https://doi.org/10.1093/envhis/emr056
Porto-Gonçalves CW (2009) De Saberes y de Territorios-diversidad y emancipación a partir de la experiencia latino-americana. Polis. Revista Latinoamericana 22(April). https://journals.openedition.org/polis/2636
Quijano A (2000) Coloniality of power, eurocentrism, and Latin America. Nepantla: Views from South 1(3):533–80.
Ramose MB (2014) Ubuntu. In: D'Alisa G, Demaria F, Kallis G (eds) Degrowth: vocaublary for a new era. Routledge, Abingdon, Oxon, pp 212–214
Richter K (2019) Struggling for another life: the ontology of degrowth (Transtext(e)s Transcultures 跨文本跨文化). J Global Cultural Stud 14(December). https://doi.org/10.4000/transtexts.1242.
Richter K (2022) Provincialising degrowth and situating Buen Vivir: a decolonial framework for the politics of degrowth. Ph.D. thesis, Goldsmiths, University of London, London, UK
Santos BS (2006) Conocer desde el Sur: Para una cultura política emancipatoria. Fondo Editorial de la Facultad de Ciencias Sociales; UNMSM; Programa de Estudios sobre Democracia y Transformación Global, Lima
Santos BS (2007) Beyond abyssal thinking: from global lines to ecologies of knowledges. Review (fernand Braudel Center) 30(1):45–89
Senplades (2007) Plan Nacional de Desarrollo 2007–2010. SENPLADES (Secretaría Nacional de Planificación y Desarrollo), Quito https://www.planificacion.gob.ec/plan-nacional-de-desarrollo-2007-2010/
Senplades (2009) Plan Nacional para el Buen Vivir 2009–2013. SENPLADES (Secretaría Nacional de Planificación y Desarrollo), Quito. https://www.planificacion.gob.ec/plan-nacional-para-el-buen-vivir-2009-2013/.
Smith LT (2012) Decolonizing methodologies: research and indigenous peoples, 2nd Revised. Zed Books Ltd, London
Strathern M (2019) A clash of ontologies? Time, law, and science in Papua New Guinea. HAU: J Ethnogr Theory 9(1):58–74. https://doi.org/10.1086/703796
Tsing, AL (2015) The Mushroom at the End of the World: On the Possibility of Life in Capitalist Ruins. Princeton, NJ: Princeton University Press
Ulloa A (2012) Producción de Conocimientos En Torno al Clima: Procesos Históricos de Exclusión/Apropiación de Saberes y Territorios de Mujeres y Pueblos Indígenas. Desigu ALdades.net Working Paper Series 21. desiguALdades.net Research Network on Interdependent Inequalities in Latin America, Berlin. https://doi.org/10.13140/RG.2.1.2519.4484
Varvarousis A (2019) Crisis, liminality and the decolonization of the social imaginary. Environ Plann e Nat Space 2(3):493–512. https://doi.org/10.1177/2514848619841809
Victor PA (2012) Growth, degrowth and climate change: a scenario analysis. Ecol Econ Econ Degrowth 84(December):206–212. https://doi.org/10.1016/j.ecolecon.2011.04.013
Videira N, Schneider F, Sekulova F, Kallis G (2014) Improving understanding on degrowth pathways: an exploratory study using collaborative causal models. Futures 55(January):58–77. https://doi.org/10.1016/j.futures.2013.11.001
Zalasiewicz J, Williams M, Haywood A, Ellis M (2011) The anthropocene: a new epoch of geological time? Phil Trans R Soc Math Phys Eng Sci 369(1938):835–841. https://doi.org/10.1098/rsta.2010.0339

Katharina Richter is Lecturer in Climate, Politics and Society at the University of Bristol. She has a PhD in Politics and has taught politics and development at UCL, Birkbeck and Goldsmiths. Her doctoral research investigated the socio-natural politics of degrowth by creating inter-epistemic dialogues with *Buen Vivir*. In 2022 and 2023, she co-lead a project on decolonizing research practices and partnerships in humanitarian assistance for the Fight Against Institutional Racism (FAIR) Network at the London School of Hygiene and Tropical Medicine. Katharina is co-organiser of Degrowth Talks, and held a previous Research Assistant post at the Reiner Lemoine Institut in Berlin.

Chapter 6
Decolonizing Nature? Worldviews of Agroecological Farmers in Germany to Address the Global Environmental Crisis

Stéphanie Eileen Domptail, Jennifer Hirsch, and Ernst-August Nuppenau

Abstract In Western Europe, farmers are embedded in a secular culture, characterized by a worldview where (hu)mans and nature are separated and opposed, capitalism rules exchanges, nature is rationally exploited, and the process of food production was long ignored. This worldview is hegemon and questioned as colonizing. Agroecological approaches and practices are said to enable farmers to entertain fundamentally different relationships with nature through their agricultural activities. Such a decolonized relationship to nature requires that farmers act based on an alternative worldview, holistic and inclusive of people and nature. Yet, we currently have little cultural information about agroecological farmers in Western Europe. We analyse narratives of four farmers to explore and document their worldviews, especially how farmers conceptualize their connection with nature. We ask how the worldview of agroecological farmers in Germany makes use of a decolonized perspective in order to reconstruct their relations to nature. Results show that both colonized and decolonized perspectives of nature co-exist but rather in the form of a struggle between worlds. Our analysis provides evidence of current cultural traits of agroecological farmers in central Germany and reveals fundamental ontological challenges in fostering the agroecological transition.

Keywords Agroecological transition · Integral cosmovision · Ontology · Cultural traits · Western paradigm

6.1 Introduction

Agriculture contributes more to the destabilization of ecosystem functions than any other single human activity (MEA 2005). Agriculture, forestry and other land-use activities accounted for 23% of the total net anthropogenic emissions of greenhouse

S. E. Domptail (✉) · E.-A. Nuppenau
Institute for Agricultural Policy and Market Research, University of Giessen, Giessen, Germany
e-mail: Stephanie.domptail@agrar.uni-giessen.de

J. Hirsch
Institute for Socio-ecological Research, Hamburger Allee 45, 60486 Frankfurt am Main, Germany

M. K. De Santo and S. E. Domptail (eds.), *Degrowth Decolonization and Development*,
https://doi.org/10.1007/978-3-031-25945-6_6

gas (IPCC Report 2019). The use of the Earth through hunting, foraging, land clearing and agriculture has always had some transformative effects on the biosphere (Ellis et al.2013). However, modern industrial agriculture has intensified and enlarged these effects spawning a global environmental crisis (Callicott 1990). Litfin (2003) and Levins (2006) refer to a mega crisis "expressed in increasing demand on depleting resources, pollution, new and resurgent diseases, climate change, growing inequality, increased vulnerability to disasters of all kinds, loss of biodiversity, the erosion of productive systems, and recurrent conflict within our species." (Levins 2006, p. 35).

At the same time, no other human activity is as basic and central for human sustenance as food production. Thus, agriculture implies both people's use of nature and our dependency on it (Sanford 2011). Agriculture represents the cornerstone of human-nature interaction where both humans' and Earth's needs may be reunited. Thereby, it has the potential to contribute greatly to solving the global environmental crises. The magnitude of the challenge may arise from its root in our culture's philosophy, as suggested by environmental sociologists.

The numerous environmental problems are the expression of a single crisis of perception (Capra 1984). The crisis "derives from the fact that we are trying to apply concepts of an outdated worldview—the mechanistic worldview of Cartesian-Newtonian science—to a reality that can no longer be understood in terms of these concepts." (Capra 1984, pp. 15–16). The mechanistic worldview of science has led to a reductionist and dualist view, where especially man is opposed to nature and is held responsible for humans' current exploitative behaviour. As suggested by Levins (2006), we focus on the crisis of *relationship*, rooted in "a pervasive and intensifying dysfunctional relationship between our species and the rest of nature". The problem is thus philosophical and requires novel thinking about our relation to nature (Pelluchon, 2021).

Decoloniality and a decolonial pluriversal perspective (Escobar 2020) do not negate Western thought per se but understand it as integral (cf. Litfin 2003; Moore 2015); nor do they "imply the absence of coloniality, but rather the ongoing serpentine movement toward possibilities of other modes of being, thinking, knowing, sensing, and living; that is, an otherwise in plural" (Mignolo and Walsh 2018, 3, p. 81). Cracks and fissures within "Western Christian civilization as spaces, places, and possibilities of and for decolonization" (Mignolo and Walsh 2018, 3, p. 81) are the focus of interest for the empirical part of this work. It scrutinizes agroecological farmers' worldviews and in particular, their current conception of nature in Germany, which implies such westernized cultural context.

Agriculture is the expression of human culture. According to Callicott (1988, p. 3), a culture's agriculture reveals its fundamental metaphysical beliefs and values. The beliefs characterizing Western Culture are "the emblematic faith in technology, the doctrine of progress, the centrality of instrumental reason, the sanctity of individual freedom, the denial of the sacred", according to Litfin (2003, p. 30). She identifies these beliefs as the cultural source of environmental destruction. These beliefs and alleged causes of our complex crisis are rooted in the same 'philosophy', which is known as the Western/secular/mechanistic worldview or the Cartesian paradigm.

Kirschenmann (2005) argues that worldviews—and today the Western worldview—shape both our perceptions and actions. Thus, worldviews are fundamental in the transformation towards a sustainable agricultural and food system.

In the science arena, environmental sociology attempts to define the fundaments of an "integrative" or ecological worldview that could reunite Earth and human needs and lead to a more sustainable human-nature relationship. The proposed worldview is rooted in systems' theory and stresses the co-evolutionary relationship between mind and action, between worldviews and the way society constructs its relationship to nature and its own structures. In parallel, agroecology proposes a set of alternative farming principles (Nicholls et al. 2017) and is presented as a paradigm for an alternative food system (Gliessman 2016; Oehen et al. 2015) in an attempt to resolve environmental and social problems. Its transformative potential (Giraldo and Rosset 2016) rests, according to De Schutter (2017), on the fact that agroecology is based on an alternative, radically different relationship of humans with(in) nature, that is, an agroecological worldview. This chapter seeks to bring together these theoretical and practical attempts to construct a worldview that sustains a decolonized relationship between people and nature. Moreover, agroecological and food justice movements demand a "shift towards a radical ontological politics beyond capital" (Moore 2015, p. 276) impending to undermine the globally dominating ontology of value and nature within the modern world-system (ibid.). Globally speaking, such political movements are still rather restrained in the Global North (Moore 2015, p. 289). In fact, cultural information about agroecological farmers in Western Europe is very scarce.

What would key concepts look like in an "agroecological" worldview that transcends the human-nature divide and inspires a more sustainable agriculture? In comparison, how does the worldview of agroecological farmers in Germany make use of a decolonized perspective in order to reconstruct their relations to nature and to the activity of production?

In its structure, the chapter first opposes the current colonizing hegemon Western Worldview to alternative theories of relating to nature, as a background supporting the following empirical analysis of actual worldviews of agroecological farmers. The next two sections present the fundament of the Western worldview—on which industrial agriculture is based—and introduce ontological principles of alternative worldviews, as potentially contributing to decolonize the relation of human with(in) nature. Section 6.4 describes the reconstructive analysis of the ontological basis for the practice of agroecology in the Western European context. Stories people tell and rely on to see the world, together with the metaphoric realms of these stories, are the expression of worldviews (Litfin 2003). Our work reconstructs the worldview of agroecological farmers and of their perceived role in the global environmental crisis by analysing transcripts of in-depth interviews with four agroecology farmers members of the Via Campesina Group of Hesse state (the Arbeitsgemeinschaft für Bäuerliche Landwirtschaft), Germany. Sections 6.5 and 6.6 show how the farmers define nature. The farmers' narratives also show the difficulty of farmers to find cultural orientation in the context of the current Western worldview they are embedded in. Through their practices and communication, the farmers daily walk the tightrope between a holistic integrative worldview and the western paradigm and economic (capitalist) system. The cultural innovations the chapter discusses include the entitlement of constructive and cooperative agency to Nature, the fundamental aim of maintaining a life basis for and beyond humans, the omnipresent systems thinking within the farm and at the landscape and global scale, as well as the role of emotions and spirituality in their interaction with or even embeddedness in nature.

Section 6.7 discusses the ontological basis for the practice of agroecology in the Western European context, and reveals fundaments for a worldview supporting the wider agroecological transition. The new metaphors of reproducing agroecosystems in co-creation with nature set in motion a cognitive transformation at the level of a whole society by raising awareness on the faulty system and by calling the modern model(s) into question (Hirsch and Norton 2012).

6.2 Background: The Emergence and Establishment of the Western Worldview in Agriculture

6.2.1 The Emergence of a Mechanistic Worldview of Nature

Capra (1984) and Merchant (1980) describe the dominant worldview in Europe in pre-modern times prior to 1500 as organic. Pre-modern science generally pursued the development of a profound understanding of nature in order to live in harmony with it. In contrast, early modern science seeks to predict and control. At the root of early modern science is the scientific revolution (Merchant 1980).

The scientific revolution was initiated with Francis Bacon in the early 17th century (Merchant 1980) and relies on three major premises. First, the scientific revolution separated objects from subjects. Natural sciences became exclusively occupied with material objects (matter) and their quantifiable and measurable properties. They excluded other properties, such as colour, sound, smell and taste, and in general life and spirituality (mind) (Capra 1984). Nature had to be observed and trialled in experiments in order to be known in absolute universal terms. Individual experience was no longer a valid approach to knowing (Mies and Shiva 2014). The belief that certainty in knowledge can be achieved only via the separation of mind and matter became central to the Western worldview and science.

Second, a great contribution to science was the analytical scientific method developed by Descartes (in the 17th century). Nature was portrayed as a machine, as "working according to mechanical laws, and everything in the material world could be explained in terms of the arrangement and movement of its parts" (Capra 1984, p. 60). Animals were *automata,* meant and programmed solely to produce. However, this analytical method has resulted in a fragmented vision of the world, coined as reductionist. The term reductionist refers to "the belief that all aspects of complex phenomena can be understood by reducing them to their constituent parts" and ignoring the role of relationships between parts (Capra 1984, p. 60). Mies and Shiva characterizes the 'scientific revolution' as 'reductionist' because it manipulates nature as inert and fragmented matter. In this way, nature loses its capacity to for 'creative regeneration and renewal' (Mies and Shiva 2014).

Finally, another important turning point in the representations of humans and nature was the incorporation in the Western worldview of Bacon's idea that "knowledge is power" and that man, through science, through knowing 'her hidden secrets', can control nature and lead/force nature to take certain paths (Merchant 2006). This contributed to the further transformation of the ancient concept of a nurturing Mother Earth into the patriarchal metaphor of the world as a machine (Capra 1984; Mies and Shiva 2014), understood and led by men for a greater general well-being (Merchant 2006).

A primary consequence of the Cartesian division between spirit and matter was that achieving an objective description of nature became the ideal of all science (Muraca 2016). Second, the goal of science became the generation of knowledge that would enable the human species to dominate and control nature and to invent technologies that could modify the world (Capra 1984, pp. 55–56; Kirschenmann 2005).

This mechanical view on nature turned into the dominant scientific paradigm for the next three centuries and shaped modern agriculture (Callicot 1990). Applied to sociology, the mechanical world view led to the idea that human minds, as powerful mechanisms, could adapt quickly to any ecological change. Catton and Dunlap (1978) termed this powerful paradigm, which underlies agrarian modernization theories in the second half of the 20th century, the 'Human Exemptionalism Paradigm'. The Scientific Method of closed experimentation, objectivity and reproducibility became the only grant of truth and knowledge. The scientific paradigm was broadened to a general Western worldview, which is summarized by Capra (1984, p. 31) in these terms:

> Belief in the scientific method as the only valid approach to knowledge; the view of the universe as a mechanical system composed of elementary material building blocks; the view of life in society as a competitive struggle for existence; and the belief in unlimited material progress to be achieved through economic and technological growth.

6.2.2 *Western Worldview and Agriculture*

Callicott (1988, 1990) summarizes how the mechanistic worldview conceptualizes modern agriculture as follows: The soil is perceived as a mere physical substrate, providing space and mechanical support for plants. Plants are complex assemblages of simple elements. The (potentially engineered) DNA of plant cells produce carbohydrates, fed as energy to "agroanimals" (as a processing step), and ultimately humans. As on an assembly line, processes are broken down to the smallest steps. The "[p]roducts are standardized; scale is magnified; and crops are specialized and monocultured." (Callicott 1988, p. 5). Importantly, agricultural goals, such as increases in yield, pest control, soil fertility, etc. (...) were believed to be achievable "Cartesian-style, by finding a separate solution for each and summing up the results" (Callicott 1990, p. 41). Callicott refers here to the reductionist approach, according to which focussing on one feature (e.g. CO_2 emissions) of a system at a time and improving it, will automatically improve the whole. (This view ignores that a change in one part of a system often creates unforeseen and hardly predictable changes in the rest of the system and does not automatically lead to an overall improvement.)

The modernization of farming systems involved a series of transitions, both physical and ontological: from labour-intensive to capital-intensive; from heterogeneous to homogeneous; from small scale to large-scale; from subjection to nature to domination of nature; from beliefs (e.g. the Maria Thün biodynamic calendar) to science (scientifically proven knowledge in the labs); and from the production of food to the production of commodities (Levins 2006, p. 38). A central example of the application of the Cartesian scientific approach to farming is the work of Justus von Liebig in the 19th century. After his identification of the role nitrogen, potassium and phosphorus play in feeding plants, von Liebig argued that the labor-intensive nutrient cycling practices could be replaced with the application of chemical fertilizers. He inspired farmers to abandon their mixed farming practices to turn towards the specialized production of a few high-value crops (Kirschenmann 2005).

The dominant narratives of conquering nature and manipulating parts of a machine—which emanated from the Scientific Revolution of early modern science in the 17th and 18th century and underlie current Western agricultural practices—were presented until recently as natural and inevitable, within both the scientific and the 'feed the world' discourse (Litfin 2003, p. 29; Sanford 2011, p. 289; e.g. the High-Level Expert Forum on how to feed the world in 2050 held in 2009). Indeed, industrialisation is both the expression and a force of enforcement of the divide between man and nature, mind and matter. Through the interaction with nature via machines, humans loose contact to the land and animals. The emotional and physical way of knowing is replaced by possibly great amounts of data resulting from smart agriculture and digital measurements further quantifying the land and animals as production units. The ensuing technology-based management sustains paternalistic (at best) relations with agricultural land and animals. It aims to control them, and so to control nature, as typical of a colonial attitude. Guzmán and Woodgate (2015) also claim that ways of farming that did not follow rules of modernity were impoverished. Indeed, in Western Europe, local knowledge has nearly disappeared and is poorly documented in scientific literature. In Africa, agroecology is being depicted as a non-modern technique and accused of maintaining the population backwards and in poverty to disqualify it (Gakpo 2020).

In this context, an increasingly loud and diverse moniroty denunciates modern agriculture as a form of colonialisation of nature. The problem is that the mechanistic worldview, with its machine metaphor and the fragmentation doctrine, is too narrow: The attempt to optimize a system by maximizing single traits ignores that trade-offs and synergies determine the behaviour of any system, and in a non-deterministic manner. In addition, machines and data transmit information on only preselected characteristics of the system, while much of the information gained through experience is lost. In the long term, it is people, thanks to centuries of relationship with agricultural soils and animals, who inform technology. Finally, the political backside of the mechanistic worldview lies in the capital-intensity of its application in modern agriculture. Capital-endorsed actors can hog agricultural resources to the detriment of actors perpetuating other ways of farming (e.g. in Brandenburg East Germany, where organic farmers built a movement: "stop land grabbing" to access land; https://buendnisjungelandwirtschaft.org/). The challenge ahead of us is to abandon the dichotomies through which we see the world today, such as the man-nature divide, the material and spiritual divide, etc... Science, as well as groups and networks from the practice in the field, are seeking new principles and paradigms for their organization.

6.3 Theoretical Developments Towards a Decolonized View of Nature and Practice

6.3.1 The Limits of Current Environmentalism

Currents of environmentalism (Guha and Martínez-Alier 1997) have developed as a response to the ecological degradation ensuing from modern instrumental usage of nature. The first current refers to exclusive conservation projects such as nature parks where nature is seen as wild, primitive, naïve, and needing (richer) human protection (from other humans) (Adams 2020; Siurua 2006). The second current refers to precision agriculture, climate-smart agriculture, nutrition agriculture and the like (Oehen et al. 2015). Climate-smart and nutrition agriculture are two examples of an attempt to improve the outcome of the agricultural activity in a reductionist manner by improving only one trait in the activity: the focus goes from yield to climate resilience or to nutritional qualities. Farmers making use of precision agriculture, often aim to limit costs through a more precise application of agrochemicals, more than any other objective (Visser et al. 2021) and intensify their technological relation to land and animals.

These hegemonic currents of environmentalism perpetuate the dichotomy between man and nature and the objectifying and instrumentalising view on nature rooted in Modern Science (Muraca 2016). They do not represent a paradigm shift, in which thoughts, perceptions, values and practices towards nature would shift fundamentally (Capra 1984). The core principles of a new paradigm shall attempt to reformulate the crucial (ethical and cognitive) questions of how we are in relationship to the environment and of how can we meet both Human and Earth needs.

A third current of environmentalism, termed Environmentalism of the Poor by Guha and Martinez-Alier (1997) refers to the struggles of small farmers, women and indigenous people to preserve their collective livelihoods as well as their vision of a self-determined and sustainable life in their community. Their language and narratives express

> a radically different [from the Western paradigm] understanding of the relation to the 'territory,' with all its inhabitants included in what can best be called a cosmo-anthropo-vision, in which interconnection among different levels of the real (biophysical, human, supernatural) leads to specific society-nature relations and nature-culture regimes. (Muraca 2016, p. 35)

Escobar (2008, p. 54) considers this a decolonial view on nature that "calls for seeing the interrelatedness of ecological, economic, and cultural processes that come to produce what humans call nature" (quoted in Muraca 2016, p. 35). Concomitantly to alternative ontologies, Hirsch and Norton (2012) propose that global issues such as climate change can best be addressed by defining a new metaphor of the world. This metaphor plays the role of a mental model. Our inner mental models mirror the outer reality. By changing our metaphors and values, we change our actions. A new metaphor can incite an individual as well as a society to act upon the outer world and shape the environment quite effectively (Capra 1984). In the case of climate change, it should allow us to "think globally, act locally". Four theoretical concepts help build up beliefs and ontologies for a decolonial worldview. These are the integral perspective (Litfin 2003), radical relationism (Muraca 2016), Gaia hypothesis (Liftin 2011) and ecofeminism, which we introduce below.

6.3.2 Principles for a Decolonized Worldview and Relation to Nature

The Integral Perspective

First, the integral perspective, developed by Karen Litfin (2003) is built on "the premise that consciousness is ontologically prior to action" (p. 29). Thus, the global problem of our relation to nature is rooted in a dysfunctional mode of consciousness. A next stage of human consciousness shall bring the understanding that humanity and nature, spirit and matter, are two dimensions of a single reality (Litfin 2003). It foresees a unique responsibility (rather than privilege) to humans to develop their consciousness and find new modes of ontologically closing the gap between mind and matter (ibid.). The integral perspective challenges the perceived **role of humans** towards nature and the scope of their responsibility while farming. Humans and specifically here farmers become accountable for shaping a new relation to nature.

Radical Relationism

Second, radical relationism proposes that "relations are ontologically prior to and constitutive of entities, rather than being conceived as external link(ing) between them" (Muraca 2016, p. 19). In other words, radical relationism argues that things acquire a meaning and become graspable entities only from the moment that one interacts with them, only *through* the interaction. Experience is created through "a web of constitutive relations that include the emotional disposition of the act of grasping itself" (Muraca 2016, p. 20). Object and subject interact and co-create one another. By contrast, the Western worldview depicts objects as pre-existing substrate and fails to acknowledge these relationships. Radical relationism even challenges the description of systems as parts *plus* their relationships, as two individually meaningful constituents of systems. One consequence of radical relationism is the necessity to consider farming systems holistically as constituted of parts, which acquire meaning **through their relationships with other parts**. For instance, soil fertility is the corner stone of healthy farming systems rather than the individual concentrations of N, P and K.

Gaia hypothesis

The Gaia theory is described by Litfin (2011) as an interdisciplinary scientific perspective that understands Earth holistically as an integrated, self-regulating biogeochemical system. As an archetype, the Gaia theory is rooted in systems' thinking and identifies three fundamental properties of living systems which can steer socio-ecological systems towards sustainability (Litfin 2011, p. 421): holism, autopoiesis and symbiotic networks, as explained in the following.

Holism is expressed in Gaia theory by the representation of Gaia (the Earth) as "the largest known instance of a living system, which in turn entails countless subsets in the form of nested living systems of biota and their environments" (Litfin 2011, p. 421). Thus, systems thinking opposes the thinking of production as a linear process from resource to waste. Holism acknowledges that the agricultural production is an act of transformation. It matters to recognize that the inputs to and outputs from production are the same environment, and thus, production is a circular transformation process. Second, **autopoietic** refers to the capacity of maintaining the system and its function over time (Meadows 2008). Thus, the purpose of the system, that is, the functions that should pertain, is a fundamental philosophical question. Currently, growth constitutes the accepted purpose of the global economy, also to deliver food. Yet, its infinite desirability on a finite planet causes Gaian-scale perturbations. Instead, the integrity and the stability of the Gaian system should become the core human purpose (Litfin 2011, p. 421). Following the Gaia hypothesis, food systems shall be managed as entities at the village, regional and global levels and within the limits of each level. The third quality of living systems, the **symbiotic networks**, refers to the idea that the world consists of relations among objects. It "stands in contrast to modern political and psychological notions of human independence" (ibid.). The Gaia theory relies on symbiotic relationships and cooperation for survival in contrast to the neo-Darwinist view of life based on competition (Tudge and Moubarac 2013). Herewith, the symbiotic perspective is revolutionary in its perspective on the nature of the interactions (competition versus cooperation) between parts within a farming system. The mental focus on reducing competition has led industrial agriculture to embrace monocultures reducing to the extreme possible interactions. The perspective of exploiting synergies and symbiotic relations among parts is currently guiding a minority of researchers and practitioners in the western world towards agroforestry, intercropping, sylvopastoralism, (...), thereby not avoiding but making use of interactions among parts of the system.

These principles appear in other emerging worldviews such as the ecological worldview, which corresponds to the vision for a scientifically informed ecological agriculture (Callicott 1990, pp. 45–46).

Eco-feminism

Eco-feminism rejects the modern Western patriarchy's special epistemological tradition of the scientific revolution due to its androcentric (men-centered) bias. Eco-feminism refuses to strip life from its spiritual and emotional dimensions. It challenges the idea of objective science and value free science to propose that research is an exchange between researcher and researchee, both influencing the outcome of the process (Mies and Shiva 2014). Eco-feminism finally questions the extent to which a system should be oriented towards production rather than reproduction, the latter thereby automatically taking the relationships with nature into account. Reproduction activities require efforts (often carried out by women) often invested in non-marketable activities in the natural and social spheres, in order to keep the system running. These are typically ignored in the western paternalistic worldview, which focusses on (marketable) outputs. In fact, the paternalistic worldview is criticised to leave the economy to capitalist exchanges (Moore 2015), which pressures farmers to produce in order to fulfil core values of productivism (Mies and Shiva 2014).

Eco-feminism also hosts a debate on the role of nature in society (Merchant 2006) and on how humans' relationship with(in) nature should be thought anew. It applies the concept of reproduction of life as central to the agricultural activity (Biesecker and Hofmeister 2010).

6.3.3 Agroecology as the Enactment of the Gaia Hypothesis?

At the farm level, agroecological approaches follow principles mimicking nature's ecological processes (Fig. 6.1).

For agriculture and the food system, the mission is to transform the farm and the food system "into a viable subset of Gaia, which means approaching it as a holistic and autopoietic living system organized as a network of relations" nourishing people and ecosystems (Litfin 2011, p. 427).

Callicott (1988, p. 3) argues that "agroecology translate[s] this abstract new vision into a concrete agricultural vocabulary: The farmstead is regarded as an artificial ecosystem with a multiplicity of diverse plant and animal constituents interacting with one another and with environing natural ecosystems in complex and mutually supporting ways". The idea that "agro-ecosystems should mimic the biodiversity levels and functioning of natural ecosystems" lies at the heart of agro-ecology while "[s]uch agricultural mimics, like their natural models, can be productive, pest resistant and nutrient conserving" (Pimbert 2015, p. 287). Principles for agroecology in the field (Nicholls et al. 2017) and in the food system (Gliessman 2016) express a systemic view of natural processes and agricultural practices.

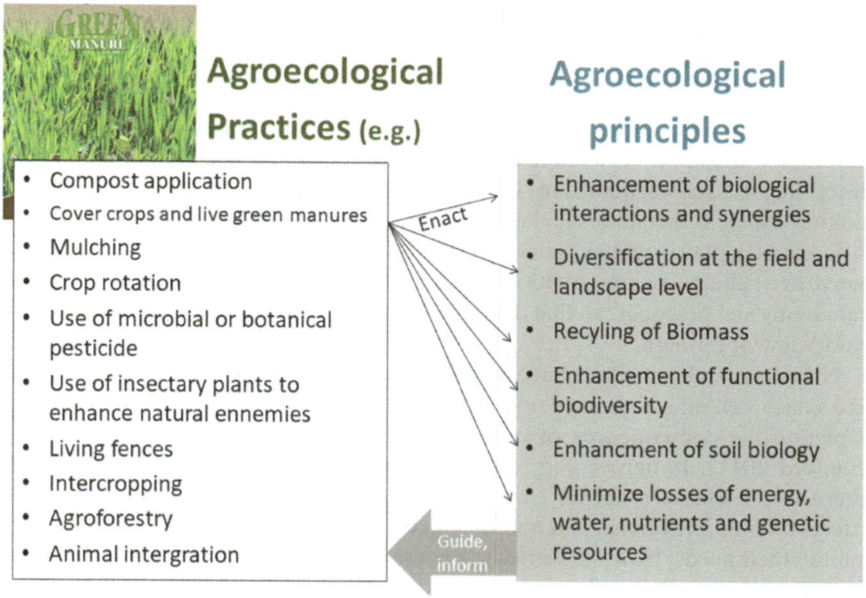

Fig. 6.1 Agroecological practices and principles (illustration after Nicholls et al. 2017; picture: www.bunnings.com.au)

In addition, and fundamental to the agroecological approach is the understanding of nature as active participant in processes of production and change (Guzmán and Woodgate 2015), reminiscent of the co-creation of subject-object mentioned earlier. Further, the material and energetic sustainability—besides the economic efficiency of agricultural systems emerges as a goal of agroecology (Guzmán and Woodgate 2015). Thus, for Callicott (1988, p. 8), agroecology expresses a new paradigm for agriculture. Litfin (2011) does suggest that organic and agroecological farmers have adopted a Gaian understanding of soil: they see and manage it as a living web of symbiotic networks rather than as a lifeless substrate for inputs and production. Rosset and Martínez-Torres (2012) suggest that this view of nature is held by the members of the agroecological movement in South America.

Yet, we are not aware of any studies exploring western farmers' worldviews and confirming this theory. The following section presents an empirical analysis of farmers' narratives presenting rare cultural evidence in this domain.

6.4 Methods and Case Study: Agricultural Narratives and Metaphors

Our exploration of farmers' worldviews is based on the analysis of their narratives. Indeed, our assumption is that language is constitutive of reality, rather than simply describing it (Escobar, quoted in Guzmán and Woodgate 2015, p. 13). This means that language such as metaphors and narratives reveal the worldview of farmers. Indeed, different metaphoric realms characterize industrial agriculture, indigenous and alternative agricultures, or conservation. This very worldview then structures agricultural paradigms and practices, so that documenting practices also reveals elements of the worldview of farmers.

Narratives and metaphors have much to do with ethics. Narratives help us develop and enact our ethical frameworks and think through various courses of action, depicting the consequences of our choices. Narratives are also morally binding (Sanford 2011). In agroecology, we expect narratives to include elements of the alternative worldviews above. They may imagine agricultural practices that consider effects on multiple human and non-human communities (Sanford 2011, p. 284) and reunify their needs. How the worldviews describe human relations with(in) nature is the focus of our investigation.

6.4.1 Qualitative Reconstructive Analysis

Reconstructive research seeks to investigate how humans mentally and thus practically shape the world they live in. Reconstruction means that we examine how social meaning is constructed by the application of linguistic means (signs and symbols) and how such social meaning is expressed in a documentary manner (cf. Kruse 2015, p. 472). We apply an integrative basis procedure (*Integratives Basisverfahren*) by Kruse (2015), which aims to let the "data talk" in an inductive manner. Yet, due to the impracticality of a solely inductive analysis, the integrative procedure recommends the use of heuristics of research objects (*forschungsgegenständliche Analyseheuristiken*) to structure and systemize the analysis openly. These heuristics can be understood as sensitizing concepts (cf. Blumer 1954) which function as 'scanners' or interpretation guides for detecting the respective structures of meaning within the text (Table 6.1).

We explore farmers' concepts of nature and the concomitant relations between the self and nature, as well as the self and the world by investigating five domains constitutive of the worldview of farmers with regards to (their) human relations with(in) nature: (1) their understanding of nature and human-nature relations; (2) the profession of farming (as the main interaction with nature) and its objectivity; (3) the nature of their knowledge on farming; (4) human responsibility towards the environmental crisis and climate change; and (5) their relation to the political and food system.

Table 6.1 Research object-specific heuristics used as scanners in the text analysis

Understanding of nature and human-nature (power) relations	How is nature described? (non-Western (Cartesian) concepts of nature, e.g. Mother Earth, a home and living being or rather a resource?)
	How does the interviewed person position him/herself in relation to nature?
	How does the interviewed person perceive he/she reunifies her needs and the needs of nature through their activity? Is there a creative interaction between human activities and nature's productivity?
Work conception, professional identity and spirituality	Does the interviewed person mention spiritual aspects? What is the role of personal relationships to things in the farming system, that is, the significance of emotion, instinct and intuition in the farming activities?
	How does the interviewed person describe his/her professional identity in relation to other farming approaches?
	Is it about competition or cooperation? (De)growth?
Knowledge and system thinking	How does the interviewed person deal with the complexity of the natural system he/she interacts with? How does he/she perceive the role of scientific knowledge and technology? How does he/she deal with uncertainty and incomplete knowledge?
	What is the aim of farm management (integrity and stability - or what other concepts guide their principles)? Does the interviewed person make use of linear thinking, or do they think in circles?
Responsibility	What is the perception of the interviewed person of his/her responsibility? At which scale (farm or world)? Do they hint at a "system" identity, such as planetary citizen or earth steward?
	How does the interviewed person perceive his/her (socio-ecological) agency with regard to the global environmental crises and especially climate change?
	How does the interviewed person describe his/her role in socio-political change? On what scale?

(continued)

Table 6.1 (continued)

Embeddedness/connectedness versus individuality, and autonomy	How does the interviewed person position him/herself in the community? In society? in the world?
	How does the person relate to the higher systems it is embedded in, especially the political and food systems? Does the person strive for autonomy?
	Is the person's agenda at the farm or local scale? or is it connected with other people's agendas and aimed at enacting change on a bigger scale?
	How is globalism perceived and what are the perceived relationships between the global and the particular, the whole and subsystems?

The information documenting the 5 heuristics all contribute to qualify the worldview of farmers with regard to the identified key characteristics of the western Cartesian worldview and existing alternatives presented in Sect. 6.4 and summarised here in Fig. 6.2.

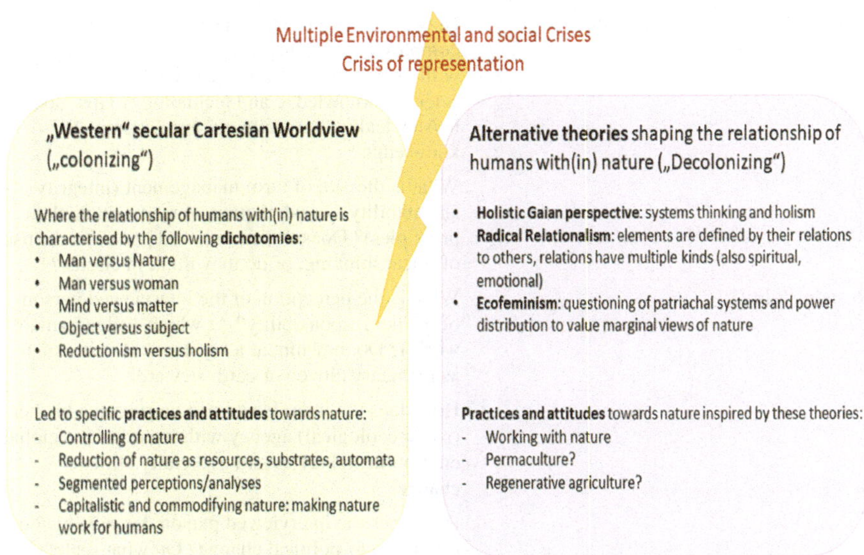

Fig. 6.2 From the Western to alternative worldviews in the face of the crisis of consciousness

6.4.2 Case Study: Sampled Agroecological Farmers in Central Germany

We chose to analyse narratives of agroecological farmers in Germany. The country has undergone a large-scale structural change in the last decades pressing conventional farmers to give up their farming or to get bigger (Domptail et al. 2018). The industrialization and increased scale of agriculture was enabled by socio-technical changes in Europe after World War II, with the need for labor in towns and the availability of nitrogen fertilizers in the whole of Western Europe. Yet, the share of organic farmers in Germany is lower than in its neighbours France and Austria, while Germany seeks to be a leader in the climate change mitigation.

The case analysed is that of a small peasant farmers' association in Western Germany in a region dominated by large-scale arable farming but adjacent to more hilly grassland landscapes. The association stands up for a socially equitable and environmentally sustainable agriculture and the establishment of conducive conditions for farming. The members of the peasant association include both conventional and organic farmers, with medium and small-sized farms. Anyone interested in supporting the aim of maintaining peasant farming, shepherds, horticulture and improve their lot may join. The association conducts political and agronomic events and supports the cooperation between farms and groups of citizens. It is also a cofounder of the European Coordination Via Campesina, an umbrella organization of peasant farmers taking part in an international fight for food sovereignty.

The second author first contacted the coordinator of a local food system group in the area. He enabled first contacts with those farmers of the association, who are strongly involved in the association's life and the local food system. At the same time, we attempted to interview both women and men farmers, as well as farmers with different farming systems such as arable, mixed, dairy system or horticulture. We left the sampling process open to conventional or organic farmers.

The sample and data consist of six in-depth interviews of which the most different four are analysed in this chapter. A first approach was made by phone, followed by an appointment with the farmers. The face-to-face interviews took place at the farmers' homesteads between October 2019 and January 2020. Farmers first narrated their farming life trajectories. The interviewer deepened some of the topics mentioned via questions developed to address the heuristics detailed above. Facts about the farms and farming systems were collected additionally and summarized in Table 6.2.

6.5 Individual Worldviews: Kaleidoscope of Motives

This section draws a picture of the striking features in the worldviews of individual farmers. Several features are then investigated more deeply in the cross-analysis in Sect. 6.6. Because our analysis uses information of both how farmers expressed themselves as well as what they say, we feel it is important in our quotes to present

Table 6.2 Characteristics of the farmers of the analysed sample

Farmer A	Farmer B
Beef production Own slaughterhouse Insect friendly mower 330 ha (incl. pasture 310 ha) 1/3 land owned	Cereals and vegetables, suckler cows Grass strips and other nature conservation measures Permaculture 90 ha (incl. 40 ha pasture) 15% land owned
Farmer C	**Farmer D**
Former scientist and employee at Bayer Dairy cows, pigs and 15 different crops Healthy diets 67 ha (incl. 40 ha pasture) 60% land owned	Former educator and religion teacher Farm Kindergarten Cattle & Pig fattening, hens and ewes 7-crops rotation 110 ha (incl. 15 ha pasture), 25% owned

the reader with the full range of the data collected and used for our interpretation: words, emphasis (capital letters), hesitations and other expressions of emotions (e.g. 'eh-eh', silences (.), !).

Farmer A. Tension between worlds

The most striking feature about Farmer A is how he appears torn between the Western Cartesian worldview with its representations of nature as pristine, and its accompanying productivist values on the one hand, and the Gaian representation of nature as a complex socio-ecosystem that humans are part of on the other hand. The tension between these worlds results in an internal struggle for the farmer, reflected in the prosodic and melodic elements of his speech, characterised by hesitation, repetition of syllables and strong intonations.

Farmer A sees himself as an active and powerful creator ("*Gestalter*") of nature and at the same time has to bend to her and adapt to her 'corset'. His agency thus alternates several times between active and passive and 'nature' is thus constructed as an antagonist, who has its own agency. Ultimately, however, he recognizes a one-sided relationship of dependence on and in favor of nature, which results from our embeddedness in it. Despite this sense of embeddedness, he and humans as such do not appear as part of nature in the sense of a harmonious symbiosis that entails a refined balance. His overall aim is to carry out the unavoidable shaping and influence by man in a way that is as gentle as possible and thus to preserve the basis of life of the human species and other non-human agents in the form of soil, air, water and biodiversity. Indeed, the 'pure form' of nature does not seem to be an option due to his awareness of living in the Anthropocene. The creation of an intact basis for life appears to overcome the two extreme forms of untouched or exploited nature. In these highly complex and biodiverse ecosystems, he is even able to recognize beauty and something wonderful. This again challenges his self-image as a rational person, as Sect. 6.6.3 details.

Farmer B: Collaboration and nature's advocate

A first essential characteristic of Farmer B's worldview may be that she sees the potential for a harmonious cohabitation on the planet earth that includes human and all sorts of non-human agents. She perceives nature as perfect, and this perfection is mainly derived from the flawlessly functioning, nested, and highly complex systems that interact on different scales. They are thereby in a fascinating balance with each other. Nature's creativity and, above all, its diversity appear as key guidelines, which she observes, tries to sustain and even facilitate further through her farming. She thus enters into a creative and equal partnership with nature (with its very own agency and qualities) and takes on the role of gardener or habitat-designer, who can provide a living environment for numerous species through her agricultural work. Consequently, she sees humans and herself as a legitimate part of nature, who can even make a positive contribution to the ecosystem. She is able to reconcile the dichotomy within the common Western perceptions of nature (untouched, wild, pristine versus over-exploited) through an agile and perilous tightrope walk between disturber and collaborator towards nature.

Farmer B's role and self-perception further derive from the inadequacy of 'the others'. She distances herself from conventional farmers who, in her opinion, are responsible for the surrounding wasteland in the landscape and the industrial agricultural desert. She positions herself as a kind of resistance fighter in advocacy of nature, who tirelessly tries to counteract the destructive tendency of other humans and particularly farmers.

> "When you LOOK at the landscape, you see (.) in fact! DESOLATEDNESS! (.) so you see insanely large areas without diversity eh (.) cultivated with one crop only (.) no hedges (.) these are things (.) I try to somehow COUNTERACT (.) so I try most of the time to CULTIVATE MIXTURES of crops (.) FLOWERIng strips surround EACH of my FIELD! HEDGES! (mhm) at the BORDER to the next FIELD (.) I try to cultivate crops that flower when NOTHING flowers in the landscape (.) ehm yes (.) I (.) actually do (.) quite some (.) nature PROTECTING (.) or nature conservation eh ya." (Farmer B, §11) [Vb1]
>
> "We have often been to the point where we think: 'this is SO much WORK, this is such A FIGHT… Why do we do that to ourselves?' But THEN again it is so IMPORTANT that we do exacTLY what we DO because ELSE no one would do it. Someone HAS to feed the EARTHworms and SOW flowers for the BEES. And this is when I think 'I actually CAN'T do this anymore', but the RESPONSIBILITY that we have with our profession is too big to abandon." (Farmer B, §17) [Vb2]

This sense of responsibility is linked to her strong understanding of ecology at the landscape scale and she sees herself as an essential provider of shelter for biodiversity in a landscape of devastation (orig. "*Ödnis*").

Farmer C: The Spiritual Scientist or the Scientific Spiritual

Farmer C's understanding of nature is characterized by a strong and above all holistic form of system thinking. This includes both, a scientific and a more emotional or spiritual component. As a PhD in natural sciences, Farmer C has a detailed understanding of interlocking natural processes and cycles down to and beyond at the molecular biological level. He carries out experiments and scientific trials within his agricultural activity and is interested in the development of innovations, which could be conducive to the maintenance, restoration or improvement (of the health) of these systems. Thereby he understands his own effectiveness and agency as very strong (frequent usage of first person action statements ('I produce', 'I make'). He is driven by his manifold visions to make the world a better place, with the superordinate and connecting goal to make of a positive contribution to the (physical and emotional) health of humans, animals and plants, and the overall system(s). His (Gaian) understanding of the world includes a strong spiritual and emotional connectedness between people, between people and animals as well as between them and their natural environment (this aspect will be explored in detail in Sect. 6.6.3). He is also striving for greater dimensions beyond the local level through better communication and the dissemination of respective knowledge (for example, through lectures and the distribution of its products abroad). We find both an indication of a strong agency, power and desire to commit to science as a validation of actions on the one side combined very interestingly with an embrace for emotional and intuitive connections with nature, feeding back to inspire innovations. This aspect is detailed in Sect. 6.6.

Farmer D: Humans, nature and stewardship

Farmer's D understanding of nature and her world view is characterized by a strong sense of system's thinking and her awareness for interdependencies on a local as well as on a global level. This sensitivity towards interdependencies goes hand in hand with an intense feeling of connectedness with her natural and social environment. Within those environments, she tries to exert as positive and harmless an influence as possible both through her farming activities as well as through her social engagement. Departing from her awareness of a momentary harmful influence of human existence on nature, her aim is to integrate her agricultural activity into natural cycles in a most compatible way. In her personal relationship with nature, she shows aspects of God-like worship that also has an awe-inspiring and fearful component. Accordingly, she attributes to nature as a superior partner its own qualities and its own agency, of which she is at the mercy as a human being. For the professional field she clearly rejects an official spirituality and thus corresponds to the classical Western values of a serious profit-oriented business. However, she assigns great importance to spiritual aspects on a personal level in the form of religious aspects, which are emphasized more strongly and presented with greater self-confidence, since they are more firmly anchored in Western Christian culture. Christian culture is deeply entrenched in German culture and goes far beyond actual visits to cult places and prayers. How the story of the genesis and religion in general frames relations to nature is largely discussed elsewhere.

6.6 Cross Analysis: Qualifying the Relationship of the Farmers to Nature

The cross-analysis deepens the main themes that emerged from the analysis of individual worldviews: these emerging themes constitute our main result as constituents of the worldview of the interviews agroecological farmers, thereby informing their relationship with(in) nature. Our analysis attempts to understand how this worldview is constructed against the existing polarized cultural background summarized in Fig. 6.2. Figure 6.3 reveals the main themes that emerged from the cross analysis and characterizing the farmers' worldview in this polarized context. The next sections explain these findingsin details.

6.6.1 From Western to Gaian Representations of Nature?

The reiterating emerging pattern across all farmers' worldviews and their relationship to nature is the perceived tension and resulting experienced struggle between the western worldview (and corresponding industrial agriculture) and engaging system thinking that resembles the Gaia representation. This tension stretches across the narratives and creates an inner struggle.

First, the narratives on the conception of nature betray a inconsistency as to the position of the farmer (of humans) in nature. This is best illustrated by the narrative of Farmer A, who first describes nature as pristine (high mountains, seas), which is "pure" and untouched, to end up at the end of the interview with an all-encompassing

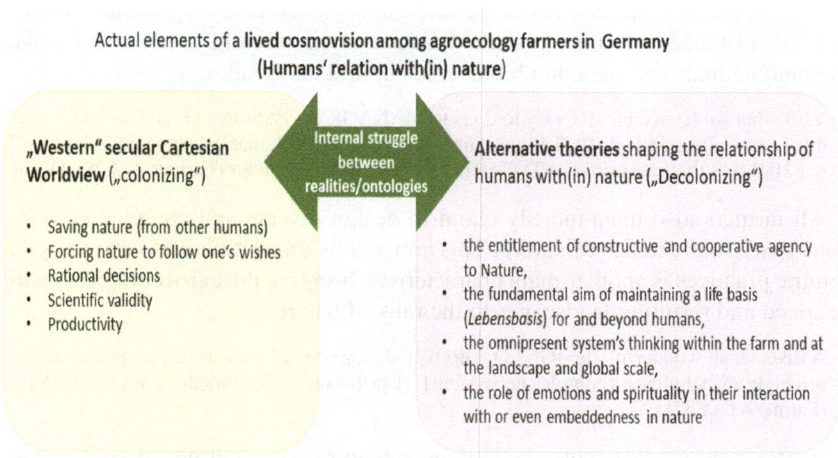

Fig. 6.3 Actual elements of a lived cosmovision among agroecology farmers in Germany, with a focus on Humans' relation with(in) nature

view of nature, nature as "all living things", with an own will, which farmers have to curve in order to make a living. Farming is perceived in this worldview as a non-natural incursion in nature, as illustrated here:

> we [humans] are the bad ones (Farmer D) [Vb3]
>
> but (.) I don't LET eh-eh-eh-nature to GROW as it in an ideal case would like to (1) but (1) Yes our lifeBASIS is eh here the INTERFERENCE in nature we have to hum.. say it this way. (Farmer A, §54). [Vb4]

The narratives show that the sampled agroecological farmers struggle intensely with positioning themselves towards or within nature: Farmers' statements suggest that they face critics from industrial or conventional farmers, which are rooted in the Western worldview and in which any farming is negative since nature is pristine. How then can the farmers justify their activity as nature-friendly, natural or ecological? This tightrope is described precisely by Farmer B:

> So: it is of course eh also a TIGHTrope walk because eh ya OTHERS SAY (.) well then let the nature completely ALONE so (.) mhm to !GROW! as IT WISHES (.) mhm but I SEE this again NOT this way because when then a plot isn't used AT ALL, then eh it is not always BETTER so you find the highest humus CONTENT [orig. AUFBAU] under PASTURES grazed by CATTLE (.) and where many different GRASSES grow and FLOWERS and all those SPEICIES they are only there when someone makes HAY and also hm MOWS (.) cuts the grass and (.) takes it AWAY so eh (1) it (.) is always about needing to maintain the EQUILIbrium. (Farmer B, §15).[Vb5]

As shown here, the interviewed farmers transcend the man-nature dichotomy to some extent by highlighting how their activities contribute to nature. Farmer D's conception of nature, for instance, is the connection and compatibility of nature conservation and the use of nature through the concept of the cultivated landscape (cf. §76). Farmers A, B and D highlight that (agroecological) farming can also improve some ecological characteristics of the landscape—and thereby legitimate their activities. For instance, it can increase the humus content of soils, create micro-habitats for small animals and maintain biodiverse cultural landscapes.

> YES what we (.) in addITION try to do is to MANAGE the PAStures (1) so that eh (.) okay this is now also political- BUT it just came to my mind YA eh that we try to build build UP our HUMUS that means to BIND CARBON Dioxyde from the air (Farmer A, §28). [Vb6]

All farmers also unanimously claim to design diverse and complex systems in their farms, to the image of nature. This aim and its translation into farm design and farming practices is another main characteristic bringing the agroecological farming practiced and resulting landscapes to the rank of nature.

> A diverse, ah wonderful, deSIGNED, GROWING together of very different species and in principle ah in this case a complex ecosystem [...], ja this for ME is something WONDERFUL (FarmerA). [Vb7]

Even when nature is described as more than pristine, cultivated landscapes, or diverse ecosystems, the narratives reveal further differences in the perception of nature as a whole: on the one hand, nature is seen as wild, powerful, cruel and in clear distinction to man.

> mhh Well NATURE can also be really CRUEL [mhm] eh when you THINK of the year-long drought in 2018. So I have the ABSOLUTE deepest RESPECT (.) and sometimes even <<laughing>> FEAR of our ehm NATUR. (Farmer D, §41) [Vb8]

On the other hand, nature is seen as processes of life (“*in each grass halm*”, Farmer B), to which all beings but also farmers contribute. In this sense, nature is a role model for designing life-support systems.

> No, rather REverence (.) nature surprises me over and over again with THINGS that I observe about her and which I perceive. (Farmer B, § 47) [Vb9]

Importantly, nature clearly has agency in these narratives: far from a substrate for farming or a resource to use, nature appears powerful and plays a role in the farming activity. In this sense, the worldview clearly departs from an extractive position to embrace the idea that nature surrounds humans. The internal struggle is to which extent human farming activities are legitimate.

6.6.2 Power Relationships Between Man and Nature: From Design to Dialogue

These different views of nature though translate into different roles of nature in farming and especially, different perceptions of the man-nature interaction. The most prominent concept to describe this interaction are the one of Design (“*Gestaltung*”).

> Well it’s not eh PURE NAture that we’re PRACticing here [mhm] it’s rather eh deSIGNING nature in order to produce FOOD. (Farmer A, §12) [Vb10]

In this example, Farmer A explains **Design**/Gestaltung as a work of sculpture, shaping somehow the wills of nature to produce foods. Nature, wild with its own will, is to some extent tamed by the farmer. The two have different wants.

Farmer B uses the same term of **Design** to explain how she works with nature to create similar complex systems. However, she sees nature as having its own functioning, which only needs to be understood in order to be better designed and strengthened:

> nature is PERfect… one must actually SEE, mhm overtake and be DIREcted by its diversity (Farmer B). [Vb11]
>
> just HOW the landscape ehm (1) just NOURISHES itself from the construction of PONDS and irrigation systems that are integrated in the landscape (.) mhm (1) that also EACH corner nearly EVERY plot has its VERY special PROPERtiesthat one should eh SEE and use. (Farmer B, §9) [Vb12]

The farmer engages in a **dialogue** with nature, where the farmer observes nature and then complements it for it to function even better (“*Simply how the landscape FEEDS simply from the construction of PONDS which one integrates in the landscape.*”). Farmer B observes that one can have nature as a **partner** because nature can solve problems itself:

> [...] GREAT how nature can already solve/regulate many things and humans actually don't need to intervene. (Farmer B) [Vb13]

This dialectic coupled with partnership characterizes especially the narratives of Farmer B and Farmer C. There are no signs of tension between their and nature's work. Nature is a partner in the game. Farmer C goes further in placing nature as an agent in his central activity: innovation. For farmer C, innovation is his "*bread and butter*"). At the same time, it is done in close observation and collaboration with nature, and more precisely with his farm plants and farm animals. Thereby, Farmer C's worldview shows no sign of the classical Western ideal of untouched nature but includes human use and creative innovations by humans in his conception of nature.

> Yes. Five years AGO I was so sassy and I PLANTED a hedge of HAZEL in a MEAdow on a SLOPE. The hazel shrubs were micronized with a MUSHroom, in fact by a TRUFFLE. [aha] because I just WANT to try out, can we actually harvest TRUFFLES in our region. I know by now they find truffles up to the south of Sweden [mhm] and why can't we try that out AS WELL? So that shows, the interACTION between NAture, between innoVAtion, und between HUMANS between ANimals is our daily BREAD. (Farmer C, §25) [Vb14]

This position represents a crescendo of the agency of nature in the farming activity to create innovations, which we term **co-creation,** here between nature as an agent and the farmer (Fig. 6.4). Co-creation appears to be the standard routine and the bread-winning activity and relationship with nature for Farmer C, who mentions it as his "bread and Butter" ("*tägliches Brot*").

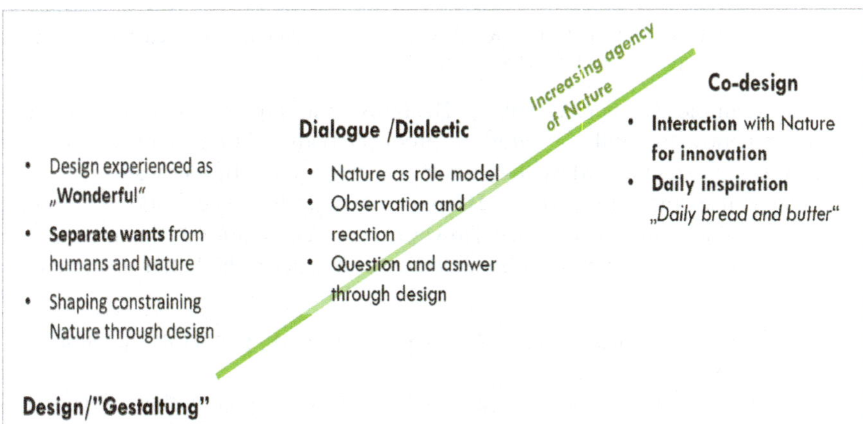

Fig. 6.4 Relationship between farmers andnaturethrough the farming activity: design, dialogue and co-creation (source: authors).

6.6.3 On the Role of the Mind: Mystic and Spiritual Relationship to Nature—The Irreconcilable?

Beyond nature's matter (ponds, humus, regulation services), the narratives betray the existence of a spiritual or even mystic component of nature (the mind). Most bonds to nature mentioned by the farmers are those with the farm animals and plants. Thus, the handling of the farm animals and plants becomes the theatre of their interaction with nature.

A respectful handling of (farm) animals is key to all Farmers. Farmer A, B and D frame their relation as that of a **stewardship**, an obligation to care for other species. Farmer B depicts her interventions (such as flower strips, late mowing, patch mowing) as nature protection ("*Naturschutz*") for wild species in an effort of conservation. Farmer A reports his great efforts to strengthen field fauna diversity:

> Someone HAS to help them! To proVIDE them with a shelter! (Farmer B) [Vb15]
>
> we are the FIRst in [OUR REGION], who have started again to (.) THIS EXTENT to MOW with the DOUBLE-blade mowing machine (.) that means that we acquired a=a machine a MOWING technique (.) which is insect-FRIEndly and hum! YES! it involved many Risks=and also !MUCH! TROUble and EFForts TIMEwise and financial hum (.) eh where we eh STILL believe today that THIS is the RIGHT thing to do. (Farmer A, §64) [Vb16]

Farmers A and D express the necessity to observe a "*mindful and respectful treatment of animals—both on a physical and emotional level*". Farmer A invested in a farm slaughterhouse to avoid traumatising animals during transportation and culling. Both frame farm (and wild) animals as "*Mitgeschöpfe*" (Beings simultaneously created literally, usual translation: God's creatures), yet with which they are not equal, but rather responsible of. This reminds of the Christian perspective of humans as highest in the hierarchy of creation, and responsible for the Garden of Eden.

Beyond this sense of responsibility, farmers did confess—albeit hesitantly and to varying extents—a more **spiritual and mystic bond** to their animals and plants. Interestingly, these statements are very hesitant in the narratives of Farmer A and D, which are more marked by the Christian worldview. Farmer A only refers to the existence of "these moments.... " but is reluctant to explicitly use the word spiritual to describe his relation with nature. Farmer C, on the contrary, openly concedes to target an intimate relationship with his animals and plants, through which he cultivates a strong emotional tie with nature. He reports from positive feelings (joy, waves), generated through the encounter with specific trees and cows. Farmer C even grants an own agency to the cows in this relationship, concurring with his perception of co-creation with nature.

> I took this a little bit as a MOTTO, that I'm in REALLY close CONTACT to the animals nowadays"; "eh we build a very TIGHT relationship (Farmer C, §9) [Vb17]
>
> [...] yes, that I go to the Pasture, that I look for body CONTACT. That I stroke them, that... the cows, when they SEE me, they also give this SIgnal back, yes [mhm] so they COME and eh actually REALLY want to PROVOKE STROKING sessions! (Farmer C, §25) [Vb18]

When a SCHOOL kid indeed also seeks CONTACT with a TREE and emBRACEs it, for me that's just the same as when I go to my animals and stroke them and I get a CERtain WAVE BACK, ja (.) of course [mhm] and I have trees in the FIELDS which I like parTICULARLY and then [mhm] eh I am HAPPY each time I'm next to them, so [mhm]. Whether it's just NONsense or not, I don't KNOW but (.) for ME personally it feels good and [yes] okay. (Farmer C, §35) [Vb19]

Nevertheless, even Farmer C is uneasy to report about this relationship and downplays it by qualifying it as possibly foolish (*orig.* "*Spinnerei*"): Indeed, Farmer A and C especially (the two men) both use vocabulary belonging to the lexical field of the productivist paradigm related to their fields of work: science and food production:

There are production methods, HIGHLY PRODUCTIVE methods, with which I can produce also INTENSIVELY and eh eh ah really ehm, I mean, a GREAT amount of FOOD. (Farmer A, §50) [Vb20]

And don't you DARE start with esoTERIC. I tell you, I have NOTHING to do with that. I Will PROVE it scientifically [mhm] (Farmer B, §1) [Vb21]

and we rePEATed the whole thing FOUR times, therefore SCIENtifically well VALIDATED [...] and to have pushed such things [innovations] and to have found out scientific trials that are SECURED (Farmer C, §11) [Vb22]

The communicated self-images intend to demonstrate rationality, a focus on productivity, growth, and science-based knowledge, values all characteristic of the classical Western worldview. Given the Western socialization of the interviewer and the interviewees, the statements appear as an effort to gain recognition in society (and from the interviewer) and to be taken seriously with their profession and agricultural approach. They reveal an inner strife between diverging elements of worldviews and conceptions of nature in the Western worldview they are embedded in, and a worldview they are creating, which encompasses—beyond matter—mind and systems.

6.6.4 "*Lebensgrundlage*": Creating a Life-Basis for One Self and the Rest of the World

The last striking concept making up the worldview of the 4 farmers with regard to nature is the one of life-basis ("*Lebensgrundlage*"). All farmers see their farming activity as actively and intentionally creating a life-basis. Three levels of life are addressed: the farmers' livelihoods, their society, and the rest of world (other species, other populations, next generations). First, farming as an interaction with nature serves as the basis for human life within the capitalist system, as the sale of produced animal and plant products generates income for the farmers and is the basis for their living.

Our lifeBASIS is eh... IS here to interFERE in natURE, we have to mhm..concede this. (Farmer A, § 54) [Vb23]

Second, the agroecological farming activity produces a life basis by providing food and nourishment for humans as well as animals. Notably, this existential meaning is already encapsulated in the German term *Lebensmittel* (literal translation: "mean for life"): Producing food, doing it well and with high nutritional quality is particularly important to Farmer C:

> I produce a FLAXSEED oil and I try to create THIS oil so that it's also VERY tasty (Farmer C,§11) [Vb24]

Third, the intervention on nature through agroecological farming is depicted to aim at the maintenance of the life basis, defined by farmer A as soil, water, biodiversity and air, to allow further life and food production - also for future generations and other species. The preservation of the environmental system's ecological reproductive capacity thereby appears as the central goal of the farming activity. This aim appears paramount to the pursuit of profit and the production of food. Thus, the protection and preservation of the life-basis are not only of existential but also of ethical relevance.

> On the other side, I think of course of our natural LIFEBASIS - which we are trying to mainTAIN and which are the BASIS for our BEING at all, which we NEED to LIVE. We have to mainTAIN our natural lifeBASIS—I mean SOILS, WATER, our AIR, our species diversity—in a pure or an Intakt state… so that we can EXIST. WE currently are BUSY desTROYING this lifebasis and WITHit we also (1) well take away the lifeBASIS of our GROWN society. (Farmer A, § 50) [Vb25]
>
> It is so IMPORTANT that we do EXACTLY what we DO, OTHERWISE no one else would do it [mhm] What they do (.) ALSO the FIELDS are only mined again. Eh yes just eh so conTINUING to ROB our LIFE-basis and eh someone HAS to feed the EARTHworms and eh SOW flowers for the BEES. And that's simply, that's when I always think, poah, no, I actually CAN'T do this any longer but actually the RESPONSIBILITY we have in our proFESSION is way too big to just give it up. (Farmer B, §17) [Vb26]
>
> So that we ehm HAVE to think very sustainably [durably] [mhm] and ehm and way BEYOND our own lives (Farmer D, §39) [Vb27]

To some extent, the concept of the life-basis manages to dissolve the dichotomy between the productive and the reproductive by making visible the interconnections and interdependencies between both spheres and by reinforcing the value of the reproductive functions that often remain invisible and taken for granted. Thereby, the Life-basis concept reconciles the views on nature as either pristine, wild and preserved versus used, tamed and exploited and thus presents a cornerstone of the agroecological farmers' cosmovision.

6.7 Discussion: Decolonizing the Relation With(in) Nature?

6.7.1 Methodology: How to Escape One's Culture

Formulating thoughts on alternative thinking is a central methodological dilemma. In this case, we, researchers from the Western tradition of thought, attempted to think beyond the boundaries we were raised and trained in. Being aware of one's own position and its consequences is demanding in itself. In addition, this philosophical challenge is strongly limited by the availability of concepts to express otherness, that is, by rhetoric. During the research process, we also had to recognize the difficulty of departing from and questioning the dichotomies between man and nature, mind and matter, modern and organic. At the same time, we needed to avoid reverting to other dualistic opposites in the attempt of answering the research question out of our own socio-cultural formed habitude: There is not *either* the Western, modern, mechanistic worldview *or* the alternative, indigenous, decolonized one to be identified from the farmers' narratives. Rather a heterogenous and multi-dimensional variety of concepts and their corresponding elements of a colonial *and/or* decolonial relational ontology prevails.

Thus, the aim was not to categorize farmers' worldviews ultimately as colonial or decolonial, thereby constructing decoloniality as a normative target that can be acquired linearly through sufficient decolonization and the elimination of differences (Mignolo and Walsh 2018, p. 81). Rather, we attempted to explore and qualify farmers' worldviews in their complexity and nuances. Our perspective aims at a process of healing integration, in the sense of a return or rediscovery of concealed and silenced elements of an original whole (beyond entities of humans and nature), which have been forcibly removed over the past five centuries (Merchant 2006).

The following discusses how the worldviews and conceptions of nature that have been reconstructed and identified from the farmers' narratives show such elements of a decolonial relational ontology

6.7.2 On Co-creation and Reproduction: Beyond Cartesian Dichotomies

Various passages and concepts in the analysed narratives point to an 'otherwise' and thus to the presence of decolonial cracks within Western civilization. Our results depict, the connection and entanglement between the agroecological farmers and nature as exceptionally plastic. We argue here that the specificity of the agroecological farmers' daily practices shaping their relations with nature consists in a radical entanglement with nature. This radical entanglement does not allow for Cartesian dualisms and the concomitant ontological categories, as Tsing Lowenhaupt (2015, p. 122) proposes (for most dichotomies, the contrast [...] suffers when it hits the ground".

Most prominently, our analysis revealed that farmers perceive and practice their relations to nature guided and enacting the concepts of design, dialogue and co-creation. These form an alternative to the two opposite poles of the Western relation to nature, i.e. conservation versus exploitation and transcend this opposition by accounting for the embeddedness in nature that the interviewed farmers perceive. Indeed, the findings point toward "a different kind of environmentalism [that] would assemble around the idea of cooperation, Gestaltung [meaning actively designing, shaping], interaction, co-creation, transformation, and ultimately inhabitation of a common, shared world" (Muraca 2016, p. 36). Thereby, the human-nature divide is increasingly transcended in a conception that recognizes the interconnections within the environment. In particular, decolonized relations between human and non-human subjects include the human-being in the web of life. This system thinking further contains an anti-capitalist element insofar as it runs counter to the isolation of entities from their relational structures (methodological individualism: Tsing et al. 2017).

Further, the concept of life-basis (*Lebensgrundlage)* introduced by the interviewees takes the same line in transcending the human-nature divide and additionally mitigates the dichotomy of the productive and the reproductive sphere, addressed in eco-feminism. By making visible the interdependencies between production and reproduction, agro-ecology goes beyond simply countering the mainstream practices. A striving for (more) egalitarian (power) relations has been identified with regard to the farmers' relationships with their animals, their plants and their soil. When cows are ascribed, their very own agency and empowered, the human-animal relationship is given a new dynamic and this goes a crucial step further towards the direction of fairness or egalitarianism. Through the attribution of non-human agency, ontologies and the hierarchy between existing categories based on anthropocentrism (i.e. first humans, then animals finally nature) are contested granting non-human actors a participation in the process of *Gestaltung* or co-creation, even if full equality cannot (yet) be achieved. Thereby, "the entangled nature of agency is brought into the foreground, as the independence of the human being from the surrounding world is deconstructed" (Redecker and Herzig 2020, p. 665).

Our findings have also shown the strong ability of the farmers to think in terms of systems: their narratives revealed a clear understanding of how their activities link to systems and supra-systems, at the local (soil health), landscape (biodiversity and small fauna conservation) and global scale (climate change, economic and social relations with the global South—the latter we did not report on here). This resonates with the conception of humans "in a flow of flows" (Moore 2015, pp. 12–13), humans as one element in a web of multiple elements as opposed to the common environmental thought of "humans control nature". At the same time, the recognition of the ever-existing entanglement of humans and nature leads the agroecological farmers (very clearly farmer B, verbatim 2) to an undeniable responsibility and obligation to take a more humble position within the system, as suggested by the integral perspective of Liftin. This responsibility is expressed in practice in our results through the up-taking of activities that aim towards reproduction in particular (soil fertility, feeding earthworms and bees, maintaining the farms) and that aim to create a life-basis for all species. Moore (2015) proposes the '*oikos'* as a concept for life-making in the production and reproduction dynamics, which would enable the process of creation/co-creation with nature.

6.7.3 *The Hardship of Cracking from Within: Internal Conflicts*

In the mids of productivism

While we presented evidence of a departure from the Cartesian dichotomies and colonial ontological categories in farmers' worldviews above, we have to acknowledge that these co-exist with evidences of concepts perpetuating the human-nature divide or suggesting power imbalances between human and non-human actors, characteristic of the Western anthropocentric worldview. This very co-existence reveals internal conflicts.

Such colonial elements can be found, among others, in associations of nature as untouched by humans (e.g. Farmer A), of humans as saviours of the world (e.g. Farmer D), or in so called "boundaries of topics": the difficulty of interviewees to address the specific topic of spirituality and emotionality and its devaluation (Sect. 6.6.3).

For one, such statements can be traced back to the socio-cultural imprint of the farmers, as they are from Germany, a western country culture. We assume that the socio-cultural and economic background of the farmers can largely be assumed to shape their relationship to nature. Although the interviewed farmers sharply criticize the relations of production and regularly break out of a neoliberal logic of efficiency (e.g. by taking on extra costs and efforts to care for the soil and animals), they remain structurally embedded in a capitalist system and ultimately have to act within its framework in order to survive. For farmer B as for all interviewed farmers, earning money and working in the cash system is natural ("natürlich"). She concedes that earning money is a bottom-line constraint: *„of course, I also have to earn MONEY with this* "(Farmer B, § 15). Somehow, the farmers cannot completely escape the way capitalism organises nature as a resource and how they can act towards it (Moore 2015). At the same, they attempt to break through or to push the relationship to nature at the far end of what the capitalist context allows by walking the tightrope described by Farmer B. This tightrope is both economic and cultural.

The farmers seek and need to conform to the Western and capitalist system in which they are embedded. In order to survive in it economically. In addition, they need to gain a sense of belonging and recognition—a basic human need—in a cultural context where productivism is a "fact". Hence, the interviewed farmers appear to experience an inner turmoil possibly stemming from the attempt as well as the need to correspond to the Western worldview outwardly. Inwardly, however, they possess a deviating view and feeling. This hints impressively at what Mignolo and Walsh (2018) refer to as the decolonial cracks and fissures that exist within the Western socio-cultural system (p. 82).

Being rational or being mad

The fracturing and fissuring of modernity's powerful colonial order are mostly accompanied by processes of struggle on superiority of narratives (in German:

Deutungshoheit, translated into English as power of directing discourses). Irrespective of whether it is a consciously and actively chosen action or not, the struggle show the power of the Cartesian mechanistic paradigm (Mignolo and Walsh 2018, pp. 82–85)

To be transparent but also clearer in our interpretations, we need to interpret this internal struggle between Western and integrative worldview elements from the cultural context of the interview. Farmers (as message-carriers) will have formulated their message under their understanding of its recipient (Kruse, 2015, p.62). The recipient was the interviewer, a younger white German female, student at the local university, that is, a western socialized non-specialist counterpart connected to a powerful western institution. Accordingly, some elements in their statements may have aimed to generate connectivity with the interviewer, that is, to gain recognition within the Western system, and avoid the risk of being devalued as acting irrationally, or as "crazy" outsiders. The somewhat "unnatural" interview situation, the awareness of being recorded and taking part in an empirical study might have led to insecurities or even nervousness, making them to revert to internalized cultural patterns of meaning out of a socio-cultural habitude (e.g. associations with nature, as mountains or natural parks, etc.).

This tendency was particularly apparent when the farmers evoked the topic of spirituality and alternative medicine, be it after a frontal question or in the course of the interviews. That the topics generated stress and the recalling to Western Cartesian separation between mind and matter shows very tangibly that spirituality is a boundary in the worldviews of farmers, a front of change. The limits of what was sayable within the Western framework of the interview situation manifested in different degrees of inner conflict and discomfort of the farmers. Together with farmers' emphasis on the importance of productivity, growth and science-based knowledge in their farming activity, which served as a sort of justification of their "otherwise", these thematic boundaries indicate the presence of what Arturo Escobar (2008) understands as coloniality, i.e. the "subalternization of local grammars and knowledge of the environment". In their everyday practice of farming, they re-exist as farmers struggle to emancipate themselves (consciously and unconsciously) from this oppression and constraints of the Cartesian mechanistic paradigm. The struggle for the freedom to decide on a different relationship to nature and worldview is accompanied by a dismissal of freedom understood in a classical Western context, i.e. human independence and superiority of nature.

The described dynamic, found in all narratives in partly different manifestations, shows an ambiguity within the worldview(s) of the farmers and further reveals the intellectual as well as social struggle that relates to a deviation from Western *worldview* and the development of a dissenting inner attitude/view of nature within the Western cultural sphere and its socio-cultural matrix.

6.8 Conclusions

Worldviews on farming and nature presenting an alternative to the dominant Western worldview and its productivist farming are essential to solving and orienting disputes concerning farming and food systems. Improving the framing of food system and addressing climate change and other environmental crises is a critical step towards more sustainability in food production. Because worldviews shape (agricultural) actions, we need more knowledge on existing views, and corresponding actions. Agroecology claims to be different from all forms of present modern agricultural systems and to have the potential to address the global environmental crisis and climate change because it is based on a different worldview; one that reflects a decolonized and holistic man-nature relationship.

The current dominant Western worldview is based on a reductionist approach to nature and follows a dual analysis of humans and nature. The underlying idea is that we can control nature. It has fostered an extractive form of modern agriculture, which alienates humans, and farmers. In particular, from an experiential communication and interaction with nature Integrative and ecological worldviews, as well as the Gaia theory seem to offer a better framework to reconcile humans' and Earth's needs. Moreover, eco-feminism offers elaborated scientific concept for reproduction, which go beyond the neoliberal worldviews. These theories address the human consciousness about their relationship to nature and they attempt to reunify mind and matter,in order to depart from a mechanistic approach to natural resource management. This reunification is coined the decolonization of nature.

Agroecological farmers claim to operate on such a decolonized relationship and contribute to a systemic understanding of nature. Our analysis reveals key cultural traits of agroecological farmers in Germany, especially with regard to their power relationship to nature. We show on the one hand very innovative mental (and practical) models of interaction between farmers and nature, built on the concepts of design, dialogue, and co-creation, as well as that of life-basis. On the other hand, we reveal a real tension within worldviews between the "Western Cartesian paradigm" and the lived relations on the farm, especially at the emotional level. The values of production and scientific rationalism seem to act as legitimizing guardrails. Our results depict innovative farmers who develop new pathways of interaction with society and nature, but who at the same time face economic and cultural pressure to fit in the "Western" world.

The chapter investigates the ontological basis for the practice of agroecology for 6 farmers playing a key role in the running and maintenance of the agroecological peasant representation in Hessen, in the Western European context taking the example of Germany. These elements may inform food science and systems to realign their relationship with nature - offering inspiration to address the global 'environmental' crisis. Indeed, it is not enough if farmers build consciousness. The lack of consciousness in the rest of society (and institutions) strongly influences farming, too. More importantly, our narratives reveal the underlying cultural tension between classical modern "western Cartesian" values and attempts to farm radically differently and so illustrates the difficulty to operate radical changes in practices. Change in culture, even in ontology, requires to engage in the current fight over the recognized and legitimate reality: whose reality(ies) will count in the future?

Acknowledgement We would like to thank the interview partners who have opened their doors to us and given us their trust. Thank you for your willingness to participate, your time, your openness, your courage, the lunches, coffees and insights into your lives. Thank you, Prof. Dr. Johanna Jacobi and Dr. Katharina Richter, for sharing your expertise with us and reviewing the book chapter. Thank you for your valuable ideas, your constructive criticism and helpful comments. Parts of this chapter are based on the thesis of Jennifer Hirsch (2022).

References

Adams WM (2020) Geographies of conservation III: Nature's spaces. In: Progress in human geography 44(4):789–801. https://doi.org/10.1177/0309132519837779

Biesecker A, Hofmeister S (2010) Focus: (Re)productivity: sustainable relations both between society and nature and between the genders. Ecol Econ 69(8):1703–1711

Blumer H (1954) What is wrong with social theory? Am Sociol Rev, 18, 3–10

Callicott JB (1988) Agroecology in context. J Agri Ethics 1:3–9

Callicott JB (1990) The metaphysical transition in farming: from the Newtonian-mechanical to the Eltonian ecological. J Agri Ethics 3(1):36–49

Capra F (1984) The turning point. Science, society, and the rising culture. Bantam Books, New York

Catton Jr WR, & Dunlap RE (1978) Environmental sociology: A new paradigm. Am Sociol, 13, 41–49

De Schutter O (2017) The political economy of food systems reform. Eur Rev Agric Econ 44(4):705–731. https://doi.org/10.1093/erae/jbx009

Domptail S, Rauer S, Müller B, Mühlleitner D, Nuppenau EA (2018) From land restructuration to land grabbing—the political context of agroecology. In: 13th European IFSA Symposium, Mediterranean Agronomic Institute of Chania, 1-5 July 2018

Ellis E, Kaplan J, Fuller D, Vavrus S, Klein Goldewijk K, Verburg P (2013) Used planet: a global history. Proc Natl Acad Sci USA 110(20):7978–7985. https://doi.org/10.1073/pnas.1217241110

Escobar A (2008) Territories of difference. Duke University Press, Durham, Place, Movements, Life, Redes

Escobar A (2020) Pluriversal politics. Duke University Press.

Gakpo JO (2020) Relying on agroecology will jeopardize Africa's food security, ag official warns. Alliance for Science. https://allianceforscience.cornell.edu/blog/2020/12/relying-on-agroecology-will-jeopardize-africas-food-security-ag-official-warns/.

Giraldo OF, Rosset PM (2016) Agroecology as a territory in dispute: between institutionality and social movements. J Peasant Stud 45(3):545–564. https://doi.org/10.1080/03066150.2017.1353496

Gliessman S (2016) Transforming food systems with agroecology. Agroecol Sustain Food Syst 40(3):187–189. https://doi.org/10.1080/21683565.2015.1130765

Guha R, Martínez-Alier J (1997) Varieties of environmentalism: essays North and South. Earthscan, London

Guzmán E, Woodgate G (2015) Transformative agroecology: foundations in agri-cultural practice, agrarian social thought and sociological theory. In: Mendez VE, Bacon CM, Cohen R, Gliessmann SR (eds) Agroecology: a transdisciplinary, participatory and action-oriented approach.

CRC Press, Boca Raton, pp 1–24. High Level Expert Forum on how to feed the world in 2050 held in 2009. 2050 high-level expert forum: the forum (fao.org).
Hirsch J (2022). Decolonizing nature? Dominant worldviews and worldviewsof agroecological farmers in Germany to address the global environmental crisis. Master's thesis. Justus Liebig University of Giessen, September 2022.
Hirsch PD, Norton BG (2012) Thinking like a planet. In: Thompson A, Bendi
IPCC (2019) Refinement to the 2006 IPCC guidelines for national greenhouse gas inventories. Subject to final copy-edit and layout prior to its final publication. https://www.ipcc.ch/site/assets/uploads/2019/06/19R_V0_01_Overview_advance.pdf
Kirschenmann FL (2005) Spirituality in agriculture. In: Leopold Center Conference Papers, 6. http://lib.dr.iastate.edu/leopold_conf/6. Accessed 27 Mai 2019
Kruse J (2015) Qualitative Interviewforschung. Ein integrativer Ansatz. BeltzJuventa, Weinheim/Basel
Levins R (2006) A whole-systems view of agriculture, people, and the rest of nature. In: Cohn A, Cook J, Fernandéz M, Reider R, Steward C (eds) Agroecology and the struggle for food sovereignty in the Americas. Russel Press, Nottingham, UK, pp 34–49
Litfin K (2003) Towards an integral perspective on world politics: secularism, sovereignty and the challenge of global ecology. J Int Stud 32(1):29–56
Litfin K (2011) Thinking like a planet: Gaian politics and the transformation of the world food system. Handbook of Global Environmental Politics, 2nd edn., pp 419–429
Mies M, Shiva V (2014) Ecofeminism. Zed Books (Critique InfluenceChange), London, New York
Meadows DH (2008) System thinking—a primer. White River Junction. Earthscan, London
Merchant C (1980) The death of nature: Women, ecology, and the scientific revolution, Harper Collins publishers, NY
Merchant C (2006) The scientific revolution and the death of nature. Isis 97:513–533
Mignolo WD, Walsh CE (2018) On decoloniality: concepts analytics praxis. Duke University Press, Durham/London
Millenium Ecosystem Assessement (2005) Ecosystems and human well-being: scenarios: findings of the scenarios working group. Island Press, Washington, DC
Moore JW (2015) Capitalism in the web of life: ecology and the accumulation of capital. Verso, London/New York
Muraca B (2016) Relational values: A Whiteheadian alternative for environmental philosophy and global environmental justice. Balkan J Philos 8(1):19–38
Nicholls CI, Altieri MA, Vazquez L (2017) Agroecological principles for the conversion of farming systems. In: Wezel A (ed) Agroecological practices for sustainable agriculture: principles, applications and making the transition. World Scientific, New Jersey/London/Singapore, pp 1–18
Oehen B, Hilbeck A, Herren H, Mueller A, Home R, Hoffman U, Nelson E, Levidow L, Pimbert M (2015) Feeding the people: agroecology for nourishing the world and transforming the agri-food system. IFOAM EU Group, Bonn. Available at: http://www.ifoam-eu.org/sites/default/files/ifoameu_policy_ffe_feedingthepeople.pdf.
Pelluchon C (2021) Ecology as new enlightenment. In: Global Solutions Journal (ed) Global Solutions Journal 2021, The World policy Forum, 7, pp 218-223
Pimbert M (2015) Agroecology as an alternative vision to conventional development and climate-smart agriculture. Development 58(2–3):286–298
Redecker S, Herzig C (2020) The peasant way of a more than radical democracy: the case of La Via Campesina. J Busi Ethics 164(4):657–670, Springer
Rosset PM, Martínez-Torres ME (2012) Rural social movements and agroecology. context, theory, and process. Ecol Soc 17(3):17. https://doi.org/10.5751/ES-05000-170317
Sanford AW (2011) Ethics, narrative, and agriculture: transforming agricultural practicethrough ecological imagination. J Agric Environ Ethics 24(3):283–303
Siurua H (2006) Nature above people: Rolston and "Fortress" conservation in the south.ethics and the environment, 11(1), 71–96. http://www.jstor.org/stable/40339115

Tsing Lowenhaupt A (2015) The mushroom at the end of the world: on the possibility of life in capitalist ruins. Princeton University Press, Princeton. https://doi.org/10.1515/9781400873548
Tsing Lowenhaupt A, Swanson HA, Gan E, Bubandt N (2017) Arts of living on a damaged planet: ghosts and monsters of the anthropocene. University of Minnesota Press, Minnesota
Visser O, Sippel SR, Thiemann L (2021) Imprecision farming? Examining the (in)accuracy and risks of digital agriculture. J Rural Stud 86:623–632. https://doi.org/10.1016/j.jrurstud.2021.07.024

Stéphanie Eileen Domptail is a senior researcher at the professorship for agricultural, food and environmental policy at the Giessen University (JLU), Germany. Stéphanie applies concepts of ecological economics and feminist economics to the study of agricultural and food systems. Stéphanie investigates agroecological systems in Germany, Nigeria and Brazil. She brings together the topics of ethics, decolonization and agroecology in their joint capacity to redesign human-nature relationships towards a more desirable state.

Jennifer Hirsch works at the institute's executive board in the department of scientific services at ISOE- Institute for Social-Ecological Research in Frankfurt, Germany. She studied American Studies and Business Administration at Goethe University Frankfurt and Nutrition Economics at Justus Liebig University Giessen. In her master's thesis at the University of Giessen, she investigated the conception of nature and the worldview of agroecological farmers in Germany in the context of the global environmental crisis using qualitative reconstructive interviews.

Ernst-August Nuppenau is retired Professor of Agricultural and Environmental Policy from the University of Giessen/Germany. He worked on the interaction of natural and human systems. An emphasis was on valuation of nature and farmers' perspective on natural resource use. A further interest has been the subject of integration of nature in farmer decision making.

Chapter 7
Aestheticizing Catastrophes?

A Comparison of the Western and Japanese Approaches to Art Creation in View of the Climate Emergency

Tamara Schneider

Abstract Without doubt, anthropogenic climate change has increased the risk of 'natural' disasters—and the impact on society will only become more severe. Art history studies societal phenomena such as culture, politics and economics through the lens of artworks, and can thus also be applied to climate-related disasters. But is there a universal response in art creation to catastrophes—particularly in intercultural comparison? In order to answer this question, in this chapter we analyse selected Japanese and US artworks dealing with recent disasters and discuss how artists' responses can help society to cope with such events. A special focus in this study lies in whether there are intercultural differences between artists' approaches to aesthetics. The study follows an iconological approach: in addition to describing artistic representations, it takes a closer look at how artworks are deeply embedded in their respective cultures. I demonstrate that while, in the wake of destruction, there is a universal urgent need for (art) creation to address and cope with the consequences, and the aesthetics of this response depend on the artist's cultural background. This has immediate implications for how art can support societal regeneration following climate-related disasters.

Keywords Aesthetics · Culture · Catastrophes · Climate change · Disasters

7.1 Introduction

Despite the ongoing struggle to reduce the impact of natural disasters on human wellbeing—whether by fighting the causes (eg, by reducing greenhouse gas emissions through climate policy) (mitigation) or, if unavoidable, by limiting the negative impacts on human beings (eg, protecting vulnerable communities from extreme weather events) (adaptation)—the recent proliferation of man-made global crises raises questions as to our capacity to effectively do the former and the need to put greater focus on the latter. The Russian invasion of Ukraine could not be

T. Schneider (✉)
Doshisha University in Kyoto, Kyoto, Japan
e-mail: tamaraschneider@gmx.de

M. K. De Santo and S. E. Domptail (eds.), *Degrowth Decolonization and Development*,
https://doi.org/10.1007/978-3-031-25945-6_7

prevented by concepts such as 'change through trade'; the COVID-19 pandemic could not be stopped by national isolation; and global warming continues despite the Kyoto Protocol and the Paris Agreement. Without neglecting the importance of early preparatory action against the multiple global threats that we face, strengthening resilience and increasing our capacity to cope with disaster appear of utmost importance for human survival.

Despite the current dominance of other crises in the public debate, recent studies have proven that the consequences of climate change with respect to mortality rates will be more dramatic than, for example, those of the COVID-19 pandemic (Carleton et al. 2021). Scientists have long warned—and continue to warn—those extreme weather events and climate-related catastrophes such as floods, droughts, storms and wildfires will become more frequent, and the negative impacts on humankind more severe. Moreover, the resulting changes to both ecosystems and societies will exacerbate existing global problems such as conflict over resources and the spread of disease (IPCC 2018: 9, Carlson 2022).

Many recent extreme weather events around the globe which have caused severe damage to societies can be attributed to climate change. In Asia, on 3 July2021, the Japanese city of Atami experienced a tragic landslide after abnormally heavy rainfall (310 mm in just 48 h) (Geospatial Information Authority of Japan 2021). Nineteen people died and some are still missing; houses were destroyed and streets were covered with debris (BBC News 2021, CNN News 2021). Just one and a half weeks later, on 14 and 15 July, Germany experienced the worst floods of the century. According to then chief executive of the Rhineland-Palatinate district of Ahrweiler, Jürgen Pföhler, this was the biggest catastrophe in the Ahr valley since World War II (WDR 2021). In North Rhine-Westphalia and Rhineland-Palatinate, 133 people died in the vicinity of the Ahr River. On 14 Julyalone, more rain fall fell in one day than usually occurs in one month. On top of the individual grief and suffering, the damage caused in the Ahr valley has been estimated at more than €3.7 billion (Teller Report 2021). This was just the latest in a series of catastrophic floods in Germany, such as the Elbe River flood in 2002 (Weichselgartner and Brévière 2011: 142). Meanwhile, in North America and Australia, the annual recurrence of wildfires and bushfires is also attributed to climate change. In California alone, these fires caused more than $10 billion in damage to properties and other structures in 2020 (ABC7 News 2020).

The above examples confirm not only the extent of the damage wrought by climate-induced extreme weather events, but also that climate change is already hitting home for those who are causing it. The long-held belief that the worst consequences of climate change would initially and primarily hit developing countries—and particularly least developed countries—has proven at least partly incorrect. Industrial nations of the Global North are already suffering from climate-related catastrophes to an extent not anticipated by experts in the early years of climate science. Hence, analysing climate-related disasters in countries of the developed Global North can provide important insights with respect to how societies cope with their effects. Even in the rich Global North, the negative impacts of global warming on societies will continue to worsen as global temperatures increase in the coming decades.

The Intergovernmental Panel on Climate Change (IPCC) predicts that the world is heading for a 3–4 °C temperature rise above pre-industrial levels by the end of this century, with devastating consequences for life on earth, if we fail to engage immediately in serious climate action (IPCC 2021: 16). It is estimated that human activities have already caused a temperature increase of more than 1 °C, and that 1.5 °C will be reached before the middle of this century (IPCC 2018: 4). Already, then, human kind faces unprecedented climate-related risks due to long-term, potentially irreversible changes to the climate system. Keeping within 1.5 °C calls for rapid, far-reaching and unprecedented changes in many aspects of society, particularly in the developed world (IPCC 2018: 15), as the G20 alone are responsible for 80% of carbon dioxide emissions worldwide. In order to limit global warming to 1.5 °C, greenhouse gas emissions must be reduced to zero by 2050 (IPCC 2018: 5).

However, even the updated nationally determined contributions of signatories to the Paris Agreement are a long way off the steep reductions needed and, if not improved, will lead to a global average temperature increase of more than 3 °C above pre-industrial levels by the end of this century (Climate Action Tracker 2022).

Hence, despite even the most ambitious mitigation efforts, some level of anthropogenic climate change is already unavoidable, which makes adaptation strategies of utmost importance. As emphasized by the IPCC, any such strategies will significantly benefit from taking cultural contexts and social values into consideration (IPCC 2014: 19). Art history as an academic humanities discipline, in turn, studies aspects of human society such as culture, politics and economics through the specific lens of artworks (Adorno 1984). By analysing artworks with its own specific methodological tools, art history can thus also contribute to understanding phenomena such as society's approach to dealing with catastrophes. While the depiction of catastrophes in art dates back to pre-modern times, contemporary art has increasingly commented on this subject. Recent examples of art exhibitions that have particularly focused on catastrophes include the Mori Art Museum Tokyo's exhibitions on *Takashi Murakami: the 500 Arhats* (Akiko 2016) and *Catastrophe and the Power of Art* (Kondo 2018). Among other things, these exhibitions raised the question of 'how art depicts such events and what role art can play in such circumstances' (Nanjo 2018, p. 5).

The earlier related exhibitions in Europe—such as the Reiss-Engelhorn Museum Mannheim's exhibition on *Mensch NaturKatastrophe. Atlantis bis heute* (Schenk et al. 2014) and the subsequent conference on *Disaster as Image: Iconographies and Media Strategies across Europe and Asia* (Juneja and Schenk 2014a)—introduced studies on the representation, mediation and interpretation of disasters in the twenty-first century. A focal point of these studies was the question: 'To what extent can we view strategies of dealing with disaster as being shaped by anthropologically constant factors?' (Juneja and Schenk 2014b: 8). As one major result, the authors agree that there is urgent need for further trans-cultural analysis on natural disasters and their imagery.

Due to its geographic and geological specifics, Japan—an island state stretching from subarctic Hokkaido to subtropical Okinawa and situated on the Pacific Ring of Fire—has a long history of coping with natural disasters, such as the Great Ansei

Earthquakes in 1855, one of the major disasters of the late Edo period. The epicentre of the earthquake was close to Edo—today's Tokyo—and the tremors and subsequent fires caused considerable damage in the Kantō region. Between7,000 and 10,000 people died.

Further examples of natural disasters in Japan include the Great Kantō Earthquake in 1923, with over 10,000 fatalities; and the Great Hanshin Earthquake in1995, which caused the deaths of over 6,400 people. All of these disasters have been represented in art—for instance, in woodblock prints, cartoons and photographs (Andreeva 2014; Weisenfeld 2014; Kirchmann 2014). One of the most prominent depictions of the 1855 earthquake and its aftermath is Kanō Kazunobu's (1816–1863) scroll series *Seven Misfortunes: Earthquake* (Nos 81–90) in the series *500 Arhats.*

Thus, Japan serves as a good case study of catastrophe representation in modern art. In comparison, the US is also prone to natural disasters—such as hurricanes, tornadoes, floods, wildfires and earthquakes– due to its vast size and diverse geology and topology; and it can also be considered as a crucible of modern Western art.

In reviews of earlier studies on the subject of catastrophe representation in Japanese contemporary art (Schneider, 2021), I have often been asked by research colleagues whether the Japanese approach can be criticized for 'aestheticizing' disasters—particularly in comparison to the more graphic catastrophe representations in most Western artworks. The answer to this question has significant implications for the aforementioned issue of how art can help societies to cope with elementary, partly unavoidable threats, and whether there are respective differences between cultures in this regard. Hence, this question deserves sound deliberation, on both theoretical and empirical grounds. What is aesthetics? Are there differences in perceptions of aesthetics between cultures? If so, what are the implications of interpreting non-Western art from a Eurocentric perspective? How do we overcome this bias; what lies beyond the Eurocentric concept of aesthetics? It becomes obvious that the term 'aestheticizing catastrophe' has pejorative connotations, suggesting suspicions that something is being concealed. There is a presumption that something is being presented in a beautiful manner in order to distance the viewer from the actual catastrophe. The scale of the danger, injustice and sorrow of the catastrophic incident, are being downplayed.

To understand artworks from different cultures, it is necessary to adopt an iconological rather than an iconographic approach. The former was first described by art historian Aby Warburg and informs humanities sub-disciplines such as cultural studies, visual studies, visual culture and Belting's visual anthropology, *Bild-Anthropologie*:

> 'Today visual culture has come to stand for an amorphous field, at the same time one which has specific connotations in different academic cultures. While in Britain and the United States visual culture (or visual studies) focuses on studying the vast variety of images available, in Germany or in France the term is closely linked to a theory or science of the image (Bildwissenschaft) or a poetics of the image, as designated in France' (Juneja and Schenk 2014b: 24).

These methods are also applicable to the analysis of Japanese art and can provide reliable insights into whether Japanese art really does aestheticize catastrophes in comparison to Western art representations. They also present an opportunity to discover a new way of dealing with art or different social interactions within the Japanese approach.

In the following sections, I will argue along three main lines. As a first step, the use of the term 'aesthetics' in art history will be explored and universalities at the crossroads between Japan and the West will be discussed on theoretical grounds. It will be shown that the adjective 'aesthetic'—even in its theoretical meaning—goes beyond the reductive and often misleading interpretation of 'beauty'.

As a second step, selected artworks by US artists depicting the effects of Hurricane Katrina will be described as a reference for comparison with Japanese artworks, and their distinctive approach to catastrophe art creation will be identified. It will be shown that even US artists take an aesthetic approach to catastrophe representation when 'aesthetics' is understood in itsmore modern sense. As a third step, selected contemporary artworks dealing with Japan's triple catastrophe in March 2011 will be analysed. It will be shown that while these works may at first glance appear 'aesthetic' in the sense of 'beautiful', a closer look reveals a distinct level of harshness which hints at a deeper meaning. The final section of the chapter summarises the findings and clarifies how theseinsightscansupport further reflections on, or give impetus to, the resolution ofcatastrophe-related grievances in society. Italso encourages a deeper understanding of differences between cultures and why it is helpful to account forthem, for example, in climate adaptation efforts.

7.2 Aesthetics in the Western Tradition and Japanese Art

So, what is 'aesthetics'? What does 'aestheticize'mean? Is the concept universal? What do we actually mean when we use this term? As there are numerous—partly ongoing—discourses in various disciplines on the subject of aesthetics (Majetschak 2019), here I will focus on the main trends immediately related to the central question of this chapter.

The term 'aesthetics'originates from ancient Greek and denotes 'perception, sensation' in the broadest sense of the meaning. It encompasses everything that moves people by touching their senses. Even as early as around the late eighth century BC, legendary ancient Greek writer Homer—who authored epic poems *The Iliad* and *The Odyssey*, the foundations of ancient Greek literature—used terms such as 'beauty' and 'harmony', albeit without any thorough theoretical analysis. For Homer, art creation was productive craftsman ship which nonetheless contained something of the divine. Socrates (469/70–399 BC) believed that beauty and good were strongly connected. A century later, Aristotle (384–322 BC) founded his own theoretical concept by examining Greek art (Aristotle, 2008). For Aristotle, art offered the ability to complete the imperfectness of nature or to model nature by mimesis– the imitative representation of the real world through art and literature. This happened through the idealisation of the real world and nature, and mainly served educational purposes. Art evokes emotions and purifies through catharsis—the process of release—thereby providing a safeguard against extreme emotions and experiences, and ultimately helping to promote better ethical behaviour. Aristotle likewise thought deeply about the connection between good and beauty. For Aristotle, beauty is an inherent quality of any object. Like Plato (c428-348 BC), Aristotle also located the nature of beauty in the formal realm: an object of beauty; harmonious proportions, a fitting or brilliant colour; an ideal design; or an artwork with a perfect arrangement of lines and colours (Sartwell 2016: 3).

This ancient Greek doctrine of aesthetics as beauty dominated the discussion on art until the nineteenth century and found its most prominent representation in Leonardo da Vinci's (1452–1519) 'Golden Ratio'. Thus, historically, aesthetics concerned principles and harmony in nature and art. This classic understanding informed the early academic discipline of art history in the West—as, for example, described and categorised in 1755 by archaeologist and librarian Johann Joachim Winckelmann. In his main work, *Anmerkungen über die Geschichte und Kunst des Alterthums*, Winckelmann suggested that only by imitating the ancient Greek masters could artists become outstanding (Winckelmann 1767). In his book, Winckelmann developed a first history of art and created a comprehensive system of Greek artworks, in which he characterized Greek sculptures by assigning them to various categories. This enabled him to describe different sculptures from across the centuries with respect to an ideal of beauty, and further made it possible for him to establish a contemporary idea of beauty. With this he founded a classification of art history based on style, whereas the lifespan of relevant artists has bee neglected. Rather, he placed greater importance on the development of form. Non-classical styles were described in nowadays well-established art historic eras like late antiquity, mannerism or baroque.

At the beginning of the twentieth century, in his book *Kunstgeschichtliche Grundbegriffe, Principles of Art History* (Wölfflin 2015), Swiss art historian Heinrich Wölfflin formulated additional criteria to delineate different epochs of art history. Through these precepts, he advanced the classification of art into distinct eras, such as Renaissance and Baroque. To this end, he used two slide projectors to make direct comparisons between different artworks. Through the establishment of a rather neutral repertoire of terms for technical features (eg, contrasting pairs such as 'linear' versus 'painterly'), Wölfflin developed a scientific basis for art history as an academic discipline. This allowed him to develop and distinguish between concrete historical categories in art, such as archaic and classical, without referring to individual images. Through this methodology, he made it possible to conduct a formal analysis of artworks within the academic discipline of art history. Eventually, a historic canon emerged consisting of the most important developments, styles and art movements—outlined most prominently in Gombrich's *The Story of Art* (Gombrich 2007).

However, by the nineteenth century, as hift away from classical principles towards a more subjective view was evident. Beauty was now thought to be in the eye of the beholder and thus to depend on the perceptions of the observer. It was understood as a positive response to something: people were delighted by, or experienced pleasure from, the artworks that they viewed. While this understanding of beauty had its early roots in the Neo-Platonism of Plotinus (204–270 AD),it became more prominent during the Age of Enlightenment. While John Locke described beauty as a purely subjective experience, David Hume and Immanuel Kant acknowledged both the objective and the subjective approach, and thus treated judgements of beauty as intersubjective (Sartwell 2016, 4). As a consequence, art and aesthetics have been understood as intuition of the artist. The artist began to express himself more and more, also using his imagination. Art movements like realism, impressionism, or expressionism occurred.

In the twentieth century, the social function of art became more prominent, as the focus shifted away from beauty towards the revolutionary potential of art—as described, for example, in Theodor Adorno's *Aesthetic Theory* (Adorno 1984), a treatise considered one of the most important contributions to the theory of aesthetics in the twentieth century in the West. In contrast toWinckelmann and Wölfflin's rather formal analysis of art, Adorno emphasized the importance of a qualitative analysis of art. To this end, he integrated socio-cultural phenomena into his analysis, exploring how art relates to society, history and politics. In this regard, he also highlighted how art can expose societal shortcomings such as economic injustice, discrimination and violence.

In today's postmodernist era, we are confronted with a whole range of new art forms—from installations to performance art and body art—which often do not fit easily with the traditional understanding of aesthetics as beauty. Artists are increasingly using radical ways to express themselves: shocking, provoking and exaggerating in order to fully exploit the revolutionary potential of art as recognized by Adorno.

One example of such radical artistic expression is Chris Burden's 1971 performance at an art gallery in Santa Ana, California (Takac 2019). The artist placed himself in front of a wall and asked his friend to shoot him in his upper left arm, turning the gallery into a visceral space of violence.

However, even this radicalisation may have reached its limits, as it seems that Western artists can no longer keep up with the radicalism of today's reality. In this regard, the documented shooting of the Russian ambassador to Turkey, Andrei Karlov, in an art gallery in Ankara in December 2016 is a telling example of how real-life extremism has supplanted artistic radicalism (Rauterberg 2017). Artists find themselves increasingly unable to disrupt and provoke society through their art, as reality is now often more radical than their artworks. This has prompted belting to proclaim the *End of Art History*. They argue that everything can now be considered as art: today, the evolution of art history is heavily influenced by market forces, such as gallery marketing campaigns, investment strategies and political trends. They further claim that no one has been able to establish a consistent history of art; that there is no such thing as universal art; and that even artists themselves are incapable of unifying art, as the variety of art is inexhaustible (Belting 2002: 24–29, 68–70 and 103).

As outlined above, in Western art, diverse new artistic epochs have emerged over the centuries. Later the market dominated and pushed this development in the West, while excluding all other less market-oriented developments (Belting 2002: 61–63). Whether the same can be said of Japanese art is questionable, especially given its particular cultural background, history and development. Japanese syncretism—apparent, for example, in the combination of Shintō and Buddhist religious concepts—allows for an easy integration of concepts from foreign countries into Japanese concepts. Therefore, Otabe (2009) has suggested that there are at least two possible ways to consider the development of Japanese art. One is from a Western perspective, with reference to the art-historical classifications of Winckelmann and Wölfflin; Otabe calls this a 'horizontal reflection'. This approach would underpin the writing of a Western-style history of Japanese art (Okakura 1903).

The second is to consider the evolution of different traditional Japanese art forms within their various schools, which Otabe calls a 'vertical reflection'. However, in 2009, during a presentation at the Philosophical Institute in Kassel University, Germany, this suggestion triggered a heated debate. Western critics argued that if we limited our analysis to individual schools—in which skills are transferred directly from masters to students and students mainly learn through imitation—no actual art development would in fact be discerned. Instead, we would encounter only aesthetization: a striving for enhanced beauty and perfection. Therefore, they claimed, Japanese art remained stuck in tradition; whereas in the West, by contrast, new art forms were free to evolve and develop.

This, however, is a somewhat reductive understanding of the learning process in Japanese art, which in fact encourages innovation once a certain level of expertise has been achieved. The individual artist benefit from centuries-long experience handed down directly by individual master artists assimilate it and then intuitively adds their own personal interpretation to the established tradition. This, in turn, creates an artistic value that is more subjective and less dependent on market value.

An additional aspect that complicates an understanding of Japanese art history—and indeed a particular peculiarity of traditional Japanese Zen art—is that it rejects any written description of techniques, structure, and composition and so on. Instead, knowledge and skills are transferred from master to student without reference to textbooks or manuals. Therefore, in this chapter, instead of referring to a (controversial) theoretical concept of Japanese art, the artworks themselves will be the focus of the analysis.

In any case, it is clear from the above discussion that the term 'aesthetics' is neither static in time nor universal across cultures. It is rather a term coined by Western art historians whose meaning has evolved over the centuries. Often, when we use the word 'aesthetic', we mean 'beautiful'; but even in the Western art historic sense, it has a deeper meaning—for example, as outlined in Adorno's *Aesthetic Theory.* In addition, given the distinct character of Japanese artworks and art history, which differ in many ways from their Western counterparts, it is misleading to apply the reductive meaning of 'aesthetics' as 'beauty' to Japanese artworks. Hence, to criticize Japanese artists for aestheticizing catastrophes based on a reductive Western understanding of 'aesthetics' is inadequate, even from a theoretical perspective.

In the following section, I will analyse selected artworks by two US artists depicting the aftermath of Hurricane Katrina to show that, under a modern understanding of aesthetics, even these artists respond to catastrophes in an aesthetic way.

7.3 US Artworks on Hurricane Katarina

This section analyses selected Western artworks depicting recent catastrophes. To this end, we will take a closer look at two responses by Western artists to the severe flooding of parts of the US Gulf Coast caused by Hurricane Katrina. Having made landfall in Florida on 25 August 2005, Katrina struck the area around New Orleans on 29 August, causing widespread flooding which lasted for weeks. Katrina and its aftermath caused about 1,800 deaths and $125 billion's worth of damage. The government's inadequate response to the catastrophe triggered intense criticism of the federal and state administrations, as well as the Federal Emergency Management Agency. The catastrophe also resulted in an artistic response, which is represented in this chapter by two deliberately chosen examples: the *Superdome Poem* by Shelton Alexander (Alexander 2015) and the *Hurricane Katrina Series* by Rolland Golden (New Orleans Museum of Art 2019). Both examples allow for the application of Adorno's interpretation of aesthetics as a means of pointing out societal shortcomings.

7.3.1 Shelton Alexander's Super dome Poem

Shelton Alexander—hailed as the 'African American Shakespeare', or 'Shake' for short—is a former US Marine who was discovered as a poet and spoken-word artist around 2001 (V-Day 2014). He also works as motivational speaker and teaches creative writing. Alexander's narrative art form can be traced back to the rich African oral storytelling tradition, in which a historian and storyteller—the '*griot*'—recounted local occurrences and historical incidents in the form of poems and songs.

In the modern society of the Global North, this storytelling tradition has largely been replaced by books and other media. However, oral storytelling conveys a high degree of authenticity due to its direct transmission from person to person. Furthermore, Alexander's poems have a catchy underlying rhythm, allowing the artist to reach a wide audience. They represent a mixed form of presentation that lies somewhere between singing and chanting. Due to their rhythmic form, Alexander's poems are also referred to as 'storytelling rap'.

In modern Western culture, this art form evolved from the Afro-American blues culture of the early nineteenth century. In addition to its emotional content, blues culture was also used to convey political messages. For instance, during the US civil rights movement of the 1960s.In the 1970s and 1980s, rap was embraced in New York hip-hop street culture, complemented by DJing, breakdancing and graffiti. Here again, emotional content, political messages and societal critiques were vital aspects of the genre. For Alexander himself, this particular style of poetry serves to release negative energy in order to help both others and oneself (Butler 2018: 4). This harks back to the Aristotelian catharsis. He further suggests that 'you can rebuild a home, but you also have to rebuild a mind, a spirit, a soul' (Alexander, without date) in order to overcome trauma.

Alexander has had first-hand experience of severe trauma himself, as a survivor of Hurricane Katrina. When the hurricane hit in August 2005, the poet—a New Orleans resident since early childhood—fled his house in a bid to escape the ensuing floods (Alexander, without date); but he got stuck in a traffic jam with other evacuees. Running low on gas, he sought shelter in the Super dome—the biggest football stadium in New Orleans, Louisiana's most populous city. The Super dome had been designated as a public 'shelter of last resort', as it was built to withstand most natural disasters. It was thus used as an emergency shelter during Hurricane Katrina, even though it had been neither designed nor tested for the task. In addition, the authorities had failed to stockpile enough supplies for emergency use and to install sufficient water supply and sanitary facilities, which led to extremely poor living conditions. After more than 30,000 people had fled from the Hurricane to the unprepared Super dome, the levees broke on 30 August; the flood soon reached the stadium and the playing field slowly filled with water. However, despite the dire conditions in the stadium and the threat from the encroaching flood, the National Guard forbade anyone from leaving. With the waters rising and no sign of rescue in sight, Shelton and 20 other flood victims decided to flee and managed to escape the Super dome. The stadium was eventually evacuated some days later by order of the Louisiana governor, again under questionable circumstances.

Superdome Poem is a two-minute video in which Alexander combines storytelling rap and personal video footage of his experiences during Hurricane Katrina. He subsequently made an additional video that is over 11 min long, entitled *I Was There: Hurricane Katrina Superdome Survivor* (Alexander, without date). In this video, he added further documentary material and described in detail his search for shelter and his eventual escape from the Super dome. Through the medium of storytelling rap, Alexander upholds the tradition of orally passing on details of real incidents to others. He makes his storytelling personal by beginning with the declaration, 'I was there' (Alexander 2015), emphasizing his status as a first-hand witness of the catastrophe. He continues by observing,'[that] which the future holds, we are unaware, storms like Katrina historically are rare' (Alexander 2015)—pointing to the reality that while such devastating weather events may have been relative rarities in the past, we cannot be so sure about the future. Indeed, these words were borne out later that year when two additional Category 5 Atlantic hurricanes hit the Gulf States—Hurricane Rita on 21–22 September and Hurricane Wilma on 19 October. In his poem, Alexander acknowledges that '[of] that destruction you bet people were scared' (Alexander 2015); and urges us to 'look but don't stare' (Alexander 2015) at the fear, despair and misery caused by Katrina, as captured in the accompanying video footage. This shows bleak images of devastation: stormy skies; flooded streets and buildings; bending trees; helpless people in rubber dinghies holding their hands over their eyes in desperation. One person clings to a lamppost to avoid being swept away by the floodwaters.

In his poem, Alexander goes on to recount the fate of those who reached the supposed safety of the Super dome. With only 'a bottle of hot water' allocated to each person, evacuees were forced to 'drink contaminated water/out of the faucet' (Alexander 2015) in the stadium toilets, which were clogged and overflowing. And while 'one MRE/Meal-Ready to Eat' (Alexander 2015)—the operational food rations for the US Armed Forces—was distributed per person, no instructions on how to prepare the food were given. As a result, 'a riot almost started' (Alexander 2015), in which Alexander had to intervene: as people 'shoved them pushed forward/over food I shouted / back up' (Alexander 2015). As a former US Marine, Alexander could not understand why the necessary supplies were not reaching the Super dome. He later stated in an interview: 'Being in the service, I know that you can get food overseas in a matter of hours or a day at most. We are right here in the city and we're about to starve to death or get dehydrated'Alexander in Butler 2018: 3–4).

The accompanying video depicts the stark contrast between the Super dome's dark interior and the glaring stadium lights. We see crowds of people—mostly African Americans—queuing for food or lying on blankets and clothing on the bare ground, with the National Guard looking on. Old and sick people sit motionless on chairs. The evacuees were expressly prevented from leaving the stadium, prompting Alexander to ask in his poem, 'Why are we been held as prisoners of war? [...] Understand, it felt more like Afghanistan' (Alexander 2015). Alexander later commented: 'Not knowing if they would just hold us in there against our will and starve us to death crossed our mind' (Alexander in Butler 2018: 4). In these bleak conditions, some of the evacuees even died.

For succour in this hopeless situation, Alexander turns to God, incorporating a traditional African American church song into his poem: 'This little light of mine / let it shine / God is good all the time' (Alexander 2015). Going on to locate his artwork within the genre of street culture, Alexander also quotes celebrated hip-hop trio Naughty by Nature, whose famous song *Everything's Gonna Be Alright* depicts the despair and poverty of growing up on the wrong side of town.

In his *Theory of Aesthetics,* Adorno wrote that '[a]rt is like a plenipotentiary of a type of praxis that is better than the prevailing praxis of society, dominated as it is by brutal self-interest' (Adorno 1984: 17, 329); and further that 'Social conflicts and class relations leave an imprint on the structure of works of art.' We see this reflected in Alexander's oral artwork and footage, through their 'intrinsic movement against society'. *Super dome Poem* starkly depicts the plight of the 30,000 mainly African Americans from poor and segregated neighbourhoods who were forced to seek shelter in the Super dome and were then held there against their will.

In sum, with hints to Adorno's *Aesthetic Theory*, Alexander's artwork clearly reflects a modern understanding of aesthetics which goes beyond the concept of pure beauty: the paradigm of ancient aesthetics that dominated art history until the early twentieth century. According to Adorno, the inadequacies of the government and ruling class can also be exposed in an aesthetic way: 'Acritical concept of society is inherent in all authentic art works' (Adorno 1984: 335). In this sense, in *Superdome Poem* Alexander criticizes the disproportionate number of African American victims of Hurricane Katrina, which can be traced back to a longstanding government policy of segregation; the resulting inability of people in poor neighbourhoods to evacuate to safer ground; and the lack of adequate help provided by government agencies (Glass 2016: 2).

7.3.2 *Rolland Golden's Katrina Series*

Similar patterns can be discerned when we examine other art works relating to Hurricane Katrina. In his response to Hurricane Katrina, Rolland Golden (1931–2019) utilizes a medium that is closer to the Western tradition of high art—that is, painting (New Orleans Museum of Art 2019). Golden was born in New Orleans and graduated from art school in the city, eventually opening his own art studio in the local French Quarter (Golden 2021). He primarily created water colors, acrylics and oil paintings in a realistic style, merged with elements of surrealism. His preferred theme was the rural Mississippi landscape, featuring barns, cow sand river scenes; but he also took a critical look at the history of Louisiana and the demolition of downtown New Orleans (La Fleur 2015, New Orleans Museum of Art 2019). Like Alexander, Golden previously served in the US Navy and even fought in the Korean War. It was not until he returned from the war that he went to art school and began painting local scenes.

Golden had a special interest both in the beauty of his surroundings and in their decay (La Fleur 2015, New Orleans Museum of Art 2019). He felt an urgent need to respond to Hurricane Katrina, with the first works in his *Katrina Series* finalized in 2007. The *Katrina Series* has since been displayed in several museums—for example, the New Orleans Museum of Art, in an exhibition entitled *Days of Terror, Month of Anguish*(Kemp 2007) (November 2007–February 2008); and most recently the Georgia Museum of Art, in an exhibition for the 15th anniversary of Hurricane Katrina entitled *Hurricane Katrina and Its Aftermath in the Art of Rolland Golden*(August–October 2020) (Georgia Museum of Art 2020). Living about 50 miles north of New Orleans, in Folsom, Louisiana, Golden himself had to withstand the storm and cope with trees falling on his house and cars (La Fleur 2015). His family lived without electricity for four days before finally making their escape to stay with relatives. Eventually, the Golden family moved to an apartment in Jackson, Mississippi, until their home was repaired in October 2005. Confronted with television footage of the vast destruction and suffering in his hometown, Golden resolved to put this on canvas even though he had decided to retire prior to Hurricane Katrina (La Fleur 2015, and New Orleans Museum of Art 2019). Once back in his hometown, he began working on his *Katrina Series*, initially through sketches, drawings and photographs of his surroundings. On 27 September, he had his first opportunity to visit New Orleans. He drove into the city on the elevated expressway and only then could see for himself the extent of the destruction underneath. On 28 September Golden took photos in the French Quarter, which still appeared deserted. Everywhere he could see tell-tale brown stains from the floodwater. The air was still heavy with the odour of decay.

Golden's *Katrina Series* consists of several paintings in oil and acrylics on canvas. He also made drawings on paper. Among other things, the paintings depict people wading through muddy water; flooded and destroyed houses; people embracing in relief after finding each other; and mud-stained graveyards. Golden categorized his *Katrina Series* as borderline surrealism, aimed at capturing on canvas the hardship and pain wrought by the disaster (Golden 2021). By transforming personal hardship into 'visual poetry' (Golden 2021), Golden tries to make it accessible to all, allowing everyone to witness and process it, and to contemplate on its possible causes and consequences (Mac Cash 2019).

One of Golden's first artistic responses to the catastrophe was his 2007 watercolour *searching for an Up-Ramp.* In this painting, we see the backs of two African Americans in hip-deep floodwater, supporting each other as they seek to escape on to dry land. Their shirts are stained brown from the waters. The person on the right stretches out their hand, either for balance or as a desperate call for help. The water is contaminated by floating debris and is so dark that we can see the reflection of the highway in the distance. On the highway stand lots of people who have already made it towards this elevated area of safety. People seem to be waving their hands in order to encourage their endangered friends in the water. Some people seem to be climbing the guard rails in order to reach dry land on the highway. 'The composition is a cross, the horizontal highway behind the vertical man and woman' (Golden 2021). This composition refers to the Christian symbol of death, rebirth and salvation.

Another revealing commentary on the humanitarian emergency is Golden's2008 acrylic painting *Hell and High Water*. According to Golden, he had seen so much and had so much to convey that it would not all fit on one canvas. Therefore, he created a triptych:

> What began as one painting grew into three, as I needed to say more about the terrible conditions these people were living in. It was a difficult road; but, when the last piece was finished, I felt purged of my feelings about the enormous tragedy (Golden 2021).

In Western iconography, the triptych is a typical devotional image or altarpiece, with themes from the life and suffering of Jesus Christ. In *Hell and High Water*, Golden created a painting focused on suffering—in this case, the suffering of the hurricane refugees of New Orleans. Again, the subjects are mainly African Americans, children, women and old people. In the panel on the left of the triptych, we see people on an elevated highway who managed to escape from the rising floodwater. Another highway above is also crowded with people, who seem to have escaped with just the clothes on their backs; only one woman carries a large brown sack with some belongings. On the highway, the refugees appear to be enduring fierce heat from the relentless sunshine. One woman wears a towel to protect her head; another protects her face with her bare hand; others are staring emptily, holding almost-naked babies or small children and half-full bottles of water. Faces express despair and lethargy. The middle panel of the triptych depicts grandparents in muddy clothes pushing their crying grandchild in a shopping cart across the highway. The shopping cart also contains two water containers. Other people are shown barefoot, suggesting they did not even have time to find their shoes before fleeing. The scorching air is smoky due to a nearby fire, with ashes flying around. A helicopter rescues people; and in the upper right corner, Golden merges in the shimmering hot, smoky air a surreal scene of flooded townhouses. The right-hand panel of the triptych shows a group of elderly, exhausted-looking women. One is holding a sign saying *Looking for L Tylor*; but the group seems to have just found the missing person and is embracing each other heartily. The face of at least one accompanying woman shows a degree of relief at this scene. Behind the group, flames are flickering, contaminating the air with smoke. On the elevated highway above, people mill around in sheer turmoil.

In sum, once again with reference to Adorno, Golden's *Katrina Series* can be interpreted as a take on aesthetics beyond the traditional concept of beauty. Adorno argues that '[a]rt is the social antithesis of society' (Adorno 1984: 11), meaning art reveals what is wrong with our supposedly social society. In the *Katrina Series*, societal grievances are exposed as Golden criticizes the inadequate attempts at rescue, especially of poorer people in segregated areas. He seems to hint that poor people were abandoned in the aftermath of the catastrophe.

7.4 Japanese Artworks on 3/11

In relation to criticisms that—particularly in comparison with the more radical and direct representations in the West—Japanese art tends to a estheticize disasters, we have shown thus far both that the theoretical understanding of aesthetics is far from unequivocal, and that Western artists also adopt aesthetic approaches in catastrophe art as the term is understood in its modern sense. In order to further scrutinize these criticisms, this section examines the artistic response to one of the biggest catastrophes in modern Japanese history: the March 2011 triple disaster.[1]

On 11 March 2011, a 9.0-magnitude earthquake hit the Tōhoku region of northern Japan, causing severe damage and triggering a devastating tsunami which hit the northern Japan prefectures of Iwate, Miyagi Fukushima, Ibaraki and Chiba. Peaking at a speed of more than 500 miles per hour and a height of 33 feet, the tsunami penetrated up to six miles inland and flooded over 500 square kilometres of Japan's Pacific Coast. The tsunami reached as far as the West Coast of North America and even—18 h after the earthquake—Antarctica, where it hit at a height of 1 foot, causing a portion of the Sulzberger Ice Shelf to break off. In addition to financial damage in the billions, the eventual toll from the earthquake and tsunami was calculated at more than 20,000 people either dead or missing, making this one of the deadliest natural disasters in Japanese history.

The third catastrophe hit when flooding from the tsunami damaged the backup generators at the Fukushima nuclear power plant, situated on the Pacific Coast. The subsequent failure of the cooling systems allowed the fuel rods to overheat; as a result, melted material burned through the containment vessel sand significant levels of radiation were released. People within a 30-mile radius of the plant were evacuated. In mid-April, the severity level of the incident reached 7—the highest on the scale

[1] This subchapter builds mainly on research published previously in Schneider (2021).

determined by the International Atomic Energy Agency—placing the Fukushima nuclear meltdown in the same category as the 1986 Chernobyl accident.

The Japanese contemporary artist Atsunobu Katagiri uses the Japanese traditional art form of *ikebana*, or flower arranging, as his chosen medium of expression. Born in 1973 in Osaka, Katagiri became head of the Misasagi *Ikebana* School in 1997. He is mainly known for his combination of traditional approaches with modern art and his collaborations with artists of different media. Katagiri specialises in large-scale flower arrangements, often working with cherry blossoms; but he is also highly skilled at producing small-scale compositions.

Ikebana (生け花), also known as *kadō* (華道)—'the way of flowers'—is the traditional Japanese art of flower arranging. The term '*ikebana*' originates from the Japanese word '*ikeru*'(生ける), meaning 'to live, become alive'; and '*hana*'(花), meaning 'flower'. '*Kadō*' uses a different kanji for 'flower'—'*ka*' (華)—and combines it with '*dō*' (道), which means 'way or path' and suggests a spiritual aspect. Thus, practising a typical Japanese art form which includes the kanji '*dō*' is considered a means of self-development (Sawano 1981: 8–9). The art form of *ikebana* can be traced back to the seventh century, when flowers were arranged as altar offerings to appease the *kami* (神)– the *Shintō* shrine deities (Sawano 1981: 16). Today, while still offered at shrines, *ikebana* is often placed in the *tokonoma* (床の間)—alcoves in private homes and teahouses. Also, in Buddhist temples (the second major religion in Japan), we still find representations of lotus flower arrangements as the immaculate holy seat of the Buddha, cast in bronze. As Buddhism has the task of preserving all living things, Buddhist priests try to prolong the life of all offered plants to the temple; and if the plants happen to grow back roots, the priests will often plant them anew (Sawano 1981: 16–17).

As a project related to 3/11, Katagiri created a series of *ikebana* which were then photographed in order to form the artwork *Sacrifice: The Ikebana of Regeneration, Offered to the Future* (2014) (Kuge 2015).On the invitation of the curator of the Fukushima Museum, Katagiri visited the closed-off zone in order to witness for himself not only the destruction caused by the triple catastrophe, but also the area's ensuing transformation. Through the flower arrangements at Minamisōma, Katagiri sought to give the area its own voice: 'From then on, I was to empty my body and mind, so that I could filter the disgrace, the grief, the sorrow, and the modest joy that shines through their cracks, and turn them into flowers' (Katagiri 2015: 99). Katagiri thus sought to empty himself in an effort to represent his surroundings by expressing the contrasting sentiments in existence at the site of the catastrophe. He also sought to reflect the changing seasons in his flower arrangements. In order to realize this concept, Katagiri stayed on the site of the catastrophe every now and then, before finally moving to Minamisōma from December 2013 until July 2014. Katagiri only worked with objects and flowers that he found locally and brought in no external materials (Katagiri 2015: 99). As soon as he found flowers suitable for an *ikebana*, he put together an arrangement and took a picture of the *ikebana* itself and the surrounding scenery. As containers for his flower arrangements, Katagiri used earthenware from the Japanese pre-historic Jōmon period (14,000–300 BC), which he had found in the Minamisōma City Museum, as well as wooden vessels, bamboo baskets and even unusual objects such as partly destroyed dolls. For Katagiri, these partly destroyed objects told stories of distant times (Akasaka 2015: 103). Along with the *ikebana* photos, Katagiri collected stories from the few people who had resisted evacuation and resettlement, and wrote down accompanying thoughts.

The photographs of Katagiri's *ikebana* can roughly be divided into three different categories. Some photos show *ikebana* backed by poetic landscape scenes. Others were taken inside destroyed buildings and reflect the remnants of vanished lives and dwellings. Still others were taken inside the Minamisōma City Museum, against a plain white or black background. The following paragraphs will analyse four examples from Katagiri's 3/11 ikebana series.

Dried silver grass (2014) is a poetic stilllife with a grass *ikebana* in a container at Migita beach, Minamisōma. The photo is divided into three horizontal parts and at first glance evokes an almost romantic peacefulness, created by the atmospheric colouring of a sunset sky in the upper two-thirds of the picture and a slim blue strip of Calm Ocean, followed by a band of wall, on which the flowerpot is placed. This type of representation could well provoke criticisms that the artist is trying to aestheticize—even beautify—the catastrophe, and to distance the viewer from the devastation and personal suffering that ensued. However, a closer look transforms the first impressions of beauty into immediately perceptible danger. The plant is not a beautiful flower, but withered silver grass. Katagiri found the grass in the expansive wilderness shortly after New Year's Eve 2014. He explicitly mentions that 'the first flower I used was not a flower, but instead a withered plant' (Katagiri 2015: 99), which can be interpreted as a reference to transience. The container– which at first sight appears to be a valuable piece of pottery, with varied colouration from the drying process– is in fact a buoy. Katagiri found the buoy– traditionally a warning and protection device– in a mountain of debris. Similarly, the wall in the lower third of the photograph on which the container is placed is a tall concrete barrier, supposed to protect the land from the ocean. While appearing peaceful in the depicted scene, the ocean can threaten lives in the event of a storm or a tsunami. Both the container and the wall are thus devices that protect humans from disasters; but in particular, the wall also disconnects us from nature. So, both the photographic artwork as a whole and the depicted *ikebana* hint at transitions—from beauty to danger; from calmness to uproar; from life to death. Beauty, serenity and life can only last for a brief moment; thereupon they fade, and death follows. The juxtaposition of these opposing impressions and the transience they evoke are comparable to the impression created by a picture puzzle– a *Vexierbild*–in which one can see two opposing things depending on the perceptions of the beholder.

At first sight, once again, *sasanqua* (2014) appears to be a beautiful baroque still life. The background is black; a camellia sasanqua blooms in vibrant red; only the skull used for the flower stand raises initial questions. For every photo that Katagiri took, there is also an underlying story which he was told by the locals. In the case of *sasanqua*, the skull originated from a two-year-old calf that had died of starvation aftermost farmers were evacuated from the contaminated area (Katagiri 2015: 100). Katagiri decorated the skull with the blossoming sasanqua, as he associated the red colour of the petals with the blood of the young calf. The farmer who provided him with the skull told Katagiri that he had likewise been ordered to kill his cows and leave the contaminated area immediately. However, the farmer resisted the order and stayed with his cows in order to bear witness to what had happened to the local people and the environment around Minamisōma. Inspired by the stories he was told and the objects he found, Katagiri later titled his work: *Sacrifice: The Ikebana of Regeneration, Offered to the Future* (Katagiri 2015: 101). In his understanding, plants sacrifice their lives to animals and people so that they can live—we all owe our lives to them, but we barely appreciate this and instead tend to destroy nature. *Ikebana*, despite its beautiful appearance at first glance, is strongly related not only to life, but also to death. Once the flowers are cut, their eventual death becomes inevitable; so the artist is obliged to keep the flowers alive for as long as possible and to create the most beautiful arrangements possible. This mindset is of utmost importance when practising the way of the flowers and is also intended to school the artist's awareness. With respect to his 3/11 photo series, Katagiri said that the aim was to 'graft a current life to a life of the past'(Katagiri 2015: 101).

A further photograph, *monochoria korsakowii*, also appears at first sight like a beautiful still life of blue-purple blooms with big green leaves skilfully arranged in earthenware from the Jōmon period against a black background. The plant– a species of the water hyacinth family in Japan, well known as *mizuaoi* (水葵)– had vanished around the swamps and lagoons of Minamisōma due to human interference with its biotope: as human settlements grew, they displaced the plant as a natural inhabitant of the area. However, on 3/11, the sea flooded the region and dragged the human settlements into the ocean, creating space for the plant's revival (Katagiri 2015: 98). As a consequence, two years after the disaster, Katagiri found the endangered lagoon-flower *mizuaoi* thriving in the area around Minamisōma and used it for his *ikebana*. Despite the humanitarian disaster and the ensuing grief, death and losses, Katagiri stated that he found 'modest joy' (Katagiri 2015: 99) in this regeneration of nature, which mirrored his admiration for self-development. The Japanese term for 'nature' is '*shizen*' (自然). However, the term '*shizen*' has a vast meaning in Japanese. The first of the two kanji, '*shi*' (自), means 'on its own terms'; and the second, '*zen*' (然), means 'yes, correct'. A more direct translation of '*shizen*' would be: 'on its own terms, therefore it exists.' In the Japanese tradition, nature is often treated as a subject; as an independent self. This means that human beings are not superior to nature but rather an equal part of it. In Katagiri's photo series, nature has the capacity to comfort, but also to replace human beings.

This becomes even more obvious in the last *ikebana* analysed here: *hydrangea*. *Hydrangea* is a photo of an *ikebana* taken in a former cinema. The focus is the cinema chairs, which Katagiri decorated with blue flowering hydrangeas. The headrests are neatly covered with embroidered white protectors. While again the photo looks calm and beautiful at first glance, with knowledge of the background, the underlying tragedy soon becomes obvious. The cinema chairs and armrests are still covered with the brown mud of the tsunami flood; and the hydrangea flower heads decorating the chairs serve as a substitute for people's heads. Again, flowers replace human beings, hinting at the equality of man and nature. The use of hydrangeas also reminds the observer of the importance of nature as an ecosystem regulator: hydrangeas—meaning 'water vessel' in Greek—are capable of absorbing excess water in an intact natural environment free from human settlements.

In sum, this section has shown that while at first sight, the response of the Japanese contemporary *ikebana* artist Katagiri to the 3/11 triple catastrophe appears to aestheticize—even beautify—this tragic event and its consequences, a deeper analysis that also reflects on the specific cultural background reveals a deeper meaning. In his works, Katagiri reflects on the transience of life; the awareness of living in the here and now; and the status of nature as a subject in itself—an animate force with equal standing to human beings. His artworks on 3/11 thus achieve far more than simply aestheticizing the disaster or even distancing people from it. In fact, they provide opportunities for both local people and global society to reflect on these matters and offer hope in times of despair.

7.5 Discussion and Conclusion

Against the background of global warming and an expected increase in extreme weather and related catastrophic events, this chapter has considered how artists can help us to cope with these partly unavoidable disasters. In particular, it has examined alleged differences between the Western and Japanese approaches to catastrophe art, with a focus on the claim that Japanese artists—particularly in comparison with their Western counterparts—tend to aestheticize disasters and thereby distance people from the respective human tragedy.

As discussed in section II, the understanding of aesthetics is primarily related to beauty originated in the ancient Greek perception of aesthetics as beauty and its evolution in Western art history from the nineteenth century onwards. However, in its modern interpretation, as emphasized by Adorno in his *Aesthetic Theory*, aesthetics also extends to the need to address social problems such as inequality, discrimination and the misuse or overuse of the natural environment. Hence, it is clear that—even from a Western theoretical perspective—there is no universal concept of 'aesthetics'.

From further comparing the Western response to Hurricane Katrina to the Japanese response to 3/11, the suggestion of a clear distinction between a Western approach that uncompromisingly and radically represents these disasters and a Japanese approach that beautifies and conceals becomes even more questionable. While Katagiri's *ikebana* photographs at first sight appear as 'aesthetic' representations, in the pejorative sense of being 'merely' beautiful, an iconological analysis reveals that these artworks also contain elements of tragedy. On the other hand, Alexander's storytelling rap and Golden's paintings on Hurricane Katrina meet the modern concept of aesthetics in their critiques of societal shortcomings.

Despite this blurring of differences with respect to the concept of aesthetics, two important distinctions should be made in this regard. First, the American artists Golden and Alexander mainly focus on people—in particular, neglected segments of society and '[a] critical concept of society is inherent in all authentic art works'

(Adorno 1984: 335)—while relying on God for help and support, due to the failure of the American government and wider society to address such problems. By contrast, Katagiri's art celebrates each living being and makes little distinction between human and non-human subjects; and there is no indication of turning to superhuman powers for help. Thus, we see two very different philosophical worldviews shining through in these artworks. The American artists share an anthropocentric worldview: the belief that human beings are the most significant living entities on Earth. This worldview originates from the Christian principle of human dominion over Earth—*dominum terra*. Genesis 1:28 says: 'God blessed the people and said to them: "Be fruitful and multiply and fill the earth and subdue it, and have dominion over the fish of the sea and over the birds of the heavens and over every living thing that moves on the earth.'The special relationship between the Christian God and humankind also allows God's most important creation to turn to Him for help in times of crisis.

The Japanese artist Katagiri, on the other hand, finds hope in the regeneration of flowers. He sought to capture the essence of Minamisōma through the flowers that he found there. In line with Japan's primary religion, Shintō, he refers to the natural circle of life, while treating nature as a subject in itself. This reflects an ecocentric worldview: a philosophy or perspective that places intrinsic value on all living organisms and their natural environment. The inherent structure reveals an acceptance of life as it is.

The second major difference lies in the addressees of calls for change and improvement. Particularly in modern and contemporary Western art, there is a strong focus on raising awareness, which follows Adorno's understanding of aesthetics. Art is thus supposed to have an outward impact on society. This trend has been reinforced in particular since the nineteenth century through the emergence of new forms of artistic expression and the iconoclastic urge to break with tradition. As a result, new artistic epochs have emerged at increasingly frequent intervals over the centuries and decades: while modern art history counts approximately distinct eras in the pre-modern period from 1000 to around 1860, the number since then has approximately doubled. Artists' heightened desire to distinguish themselves from their predecessors has also been fuelled by the increasing mercerisation of art in the modern age (Belting 2002: 64). Modern artists are also increasingly moved to exposé obvious grievances in society. Alexander alludes to this intention in his *Super dome Poem* when he says, 'a reason I share the truth is hard, at least it's fair, sad but true, I would take you, I was there' (Alexander 2015);when he exhorts us to 'look but don't stare' (Alexander 2015)at the carnage—both natural and political—wrought by the hurricane; and when he highlights the need to rebuild not only homes, but also 'minds, spirits and souls'(Alexander, without date). And Golden has similarly explained his motivation for the *Katrina Series*as follows: 'It is my hope, these paintings will help those who see them better understand what occurred and that we continue to suffer, long after Katrina died. It will happen again somewhere; let us hope we will be better prepared' (Golden 2021). Both artists thus address society as a whole and draw attention to social issues in their artworks.

In contrast, in Japanese art forms such as *kadō*, there is a focus on the schooling of awareness and self-improvement, and they are thus rather inwardly directed. Knowledge is handed down vertically in schools from teacher to student. Few written teaching materials exist. These art forms are highly respected; schools with a long tradition are valued and strive to continue that tradition in order to pass down established knowledge and skills. Only by following the school's tradition can students fully comprehend and master the techniques and '*dō*', as self-development for deepening their understanding of life (Sawano 1981: 8). These art forms serve the market much less than their Western counters parts. This is evident even in the materials used—for example, in *ikebana*. 'A branch or flower may be chosen not for its perfection, but for its imperfection' (Sawano 1981: 11)– such as the withered plant in Katagiri's art work *dried silver grass*. 'Such things remind us of the fragility of all living things' (Sawano 1981: 11). On this point, Katagiri commented on his *ikebana* photo series *Sacrifice: The Ikebana of Regeneration, Offered to the Future:* 'I even sensed beauty in the raging nature, in a presence that was beyond human reason. The power of nature was so overwhelming that I even forgot that many human lives were being lost in its midst' (Katagiri 2015: 98). He continued: 'From then on, I had to empty my body and mind, so that I could filter the disgrace, the grief, the sorrow, and the modest joy that shines through their cracks, and turn them into flowers' (Katagiri 2015: 98).

In sum, claims that Japanese artists aestheticize catastrophes—through a simplistic interpretation of beauty—perpetuate a rather old-fashioned Eurocentric worldview; ignore the nuances of Japanese artistic traditions and culture; and create barriers to mutual understanding and efforts to cope cooperatively with unavoidable catastrophes. By contrast, understanding and accepting other realities beyond the narrow Western frame of interpretation could lead to significant progress in human development and promote greater coexistence with nature. It could also heighten awareness of anthropogenic threats to nature, promoting a precautionary approach to our treatment of the environment and encouraging us to live in harmony with our natural surroundings. Japanese traditional art forms such as *kadō* in particular teach this awareness. On the other hand, Japanese art forms also teach us how to overcome unavoidable hardship by accepting the circle of life as it is.

Therefore, Japanese art and the processes through which it is produced can be understood as means of both *ex ante* anticipating and preventing harm and *ex post* accepting and coping with unavoidable damages– a way of thinking that could be highly valuable for climate mitigation and adaptation.

In order to further scrutinize both the differences and similarities in Western and Japanese disaster art, future studies should broaden the empirical basis by analysing a greater number of respective Western artworks from different countries, genres and art eras. The same applies to the study of Japanese catastrophe-related art. Such a humanities-based research programme could prove a valuable complement to the rather technical economics, engineering and natural science-focused research on, and development of, effective responses to disasters– particularly in the wake of anthropogenic global warming and the expected increase in climate-related extreme weather events.

References

ABC7 News (2020) Wildfire. Damage from California's wildfires are estimated at $10 billion, experts say by David Louie. Available via https://abc7news.com/california-wildfires-cost-of-cal-fire-stanford-wildfire-research/6897462/. Accessed 31 Jan 2022

Adorno T (1984) Aesthetic theory. Routledge, London, New York

Akasaka N (2015) The sacrifice of a flower. In: Kuge M (ed) Sacrifice—TheIkebana of regeneration, offered to the future.Seigensha, Kyoto, pp 102–104

Akiko M et al (eds) (2016) Takashi Murakami—the 500 arhats. Heibonsha, Tokyo

Alexander S (without date) I was there—Hurricane Katrina Superdome Survivor. Available via https://www.history.com/topics/natural-disasters-and-environment/i-was-there-hurricane-katrina-superdome-survivor-video. Accessed 5 May 2022

Alexander S (2015) Hurricane Katrina—Superdome Poem. Available via https://www.youtube.com/watch?v=jzv4L1ggfnMAccessed. Accessed 5 May 2022

Andreeva A (2014) The "Earthquake Insect"—conceptualizing disasters in pre-modern Japan. In: Juneja M, Schenk GJ (eds) Disaster as image—iconographies and media strategies across Europe and Asia. Schnell & Steiner, Regensburg, pp 81–90

Aristotle (2008) Poetics. Cosimo Classics, New York

Belting H (2002) Das Ende der Kunstgeschichte—Eine Revision nach zehn Jahren. CH Beck, Munich

BBC News (2021) Japan landslide: 20 people missing in Atami city. Available via https://www.bbc.com/news/world-asia-57704967. Accessed 8 Nov 2021

Butler J (2018) Shelton Alexander—Surviving in the Superdome During Hurricane Katrina. Available viaLWOS. https://lastwordonsports.com/nfl/2018/04/21/shelton-alexander-surviving-superdome-hurricane-katrina/. Accessed 5 May 2022

Carleton TA et al (2021) Valuing the global mortality consequences of climate change accounting for adaptation costs and benefits. National Bureau of Economic Research, Cambridge Mass (Working Paper 27599)

Carlson CJ et al (2022) Climate change increases cross-species viral transmission risk. Nature. https://doi.org/10.1038/s41586-022-04788-w

Climate Action Tracker (2022) CAT Emissions Gap. Available via https://climateactiontracker.org/global/cat-emissions-gaps/. Accessed 7 Apr 2022

CNN News (2021) About 20 people missing and 2 dead after mudslide wipes out homes in Japan's Atami City. Available https://edition.cnn.com/2021b/07/03/asia/japan-mudslide-atami-intl-hnk/index.html. Accessed 8 Nov 2021

Georgia Museum of Art (2020) Hurricane Katrina and its aftermath in the art of Rolland Golden. Available via https://georgiamuseum.org/exhibit/rolland-golden/. Accessed 7 Jun 2022

Geospatial Information Authority of Japan (2021) 令和3年(2021年)7月1日からの大雨に関する情報. Available via https://www.gsi.go.jp/BOUSAI/R3_0701_heavyrain.html. Accessed 8 Nov 2021

Glass V (2016) Underpinnings of government policy and Hurricane Katrina. Available via Process.https://www.processjmus.org/underpinningskatrina. Accessed 5 May 2022

Golden R (2021) The world of Rolland Golden. Available via https://www.rollandgolden.com/. Accessed 9 May 2022

Gombrich EH (2007) The story of art. Phaidon, London, New York

IPCC (2014) Climate change 2014 synthesis report—summary for policymakers. IPCC, Geneva

IPCC (2018) Global warming of 1.5 °C—summary for policymakers. IPCC, Geneva

IPCC (2021) Climate change 2021—the physical science basis—summary for policymakers. IPCC, Geneva

Juneja M, Schenk GJ (eds) (2014a) Disaster as Image. In: Iconographies and media strategies across Europe and Asia. Schnell & Steiner, Regensburg

Juneja M, Schenk GJ (2014b) Viewing disasters—myth, history, iconography and media across Europe and Asia. In: Juneja M, SchenkG. (eds) Disaster as Image—Iconographies and Media Strategies across Europe and Asia. Schnell & Steiner, Regensburg, pp 7–40
Katagiri A (2015) Man and Flowers. In: Kuge M (ed) Sacrifice–the Ikebana of regeneration, offered to the Future. Seigensha, Kyoto, pp 98–101
Kemp JR (2007) Katrina–days of terror, month of Anguish. In: The New Orleans Museum of Art (ed): Katrina–days of terror, month of Anguish–paintings by Rolland Golden. New Orleans Museum of Art, New Orleans, pp 9–17
Kirchmann K (2014) Constructions of otherness. Images of pain, suffering and stoicism during Japanese disaster 2011. In: Juneja M, Schenk GJ (eds) Disaster as image–iconographies and media strategies across Europe and Asia. Schnell & Steiner, Regensburg, pp 145–152
Kuge M (ed) (2015) Sacrifice–the Ikebana of regeneration, offered to the Future. Seigensha, Kyoto
La Fleur D et al (2015) Rolland Golden La Legends Interview & Short Documentary 2015. Available via Louisiana Educational Television Authority. https://www.youtube.com/watch?v=xkRoXoCcyMw. Accessed 9 May 2022
Mac Cash D (2019) Louisiana Painter Rolland Golden dies at 87; artist went from French Quarter bohemian to Katrina chronicler.Available via https://www.nola.com/entertainment_life/arts/article_2519be14-9c58-11e9-a3a2-d7216f386e00.html. Accessed 10 May 2022
Majetschak S (2019) Ästhetikzur Einführung. Junius, Hamburg
Nanjo F (2018) Foreword. In: Kondo K et al (eds) Catastrophe and the power of art. Heibonsha, Tokyo, p 5
New Orleans Museum of Art (2019) Remembering Rolland Golden, a Louisiana artist who captured both natural splendor and devastation. Available via https://noma.org/remembering-rolland-golden-a-louisiana-artist-who-captured-both-natural-splendor-and-devastation/. Accessed 9 May 2022
Okakura K (1903) The ideals of the East. Cover, London
Otabe T (2009) Ästhetik im kulturellen Austausch zwischen Japan und Deutschland. Drei Stufen der Globalisierung. Ein Beitrag zur interkulturellen Ästhetik. Paper presented at Institut für Philosophie, Kassel Unviersity, Germany, 28 June 2009
Rauterberg H (2017) Im Kugelhagel Der Realität. Die Zeit 3:12
Sartwell C (2016) Beauty. Available via The Stanford Encyclopedia of Philosophy. https://plato.stanford.edu/archives/win2017/entries/beauty/. Accessed 6 Aug 2019
Sawano T (1981) Ikebana–basic principles. Ward Lock Limited, London
Schenk GJ et al (eds) (2014) Mensch. Natur. Katastrophe–Von Atlantis bis heute. Schneller & Steiner, Regensburg
Schneider T (2021) Transitions after destruction–the artistic response to the 2011 Triple Catastrophe in Japan.In: Krings L et al (eds) Transitions–crossing boundaries in Japanese Philosophy. Chisokudō, Nagoya, pp 316–338
Takac B (2019) When Chris Burden tried to shoot himself for the Sake of Art. Available via Widewalls Magazine. https://www.widewalls.ch/magazine/chris-burden-shoot. Accessed 6 Jul 2022
Teller Report (2021) Preliminary estimate–Ahrweiler district puts flood damage at more than 3.7 billion. Available via https://www.tellerreport.com/business/2021-08-07-preliminary-estimate--ahrweiler-district-puts-flood-damage-at-more-than-3-7-billion.H17Wfdzh1K.html. Accessed 6 May 2022
V-Day (2014) Shelton Alexander. Available via Anniversary Events.https://www.vday.org/anniversary-events/superlove/bios/alexander.html. Accessed 5 May 2022
WDR (2021) Ahrtal unter Wasser–Chronik einer Katastrophe. Available via https://reportage.wdr.de/chronik-ahrtal-hochwasser-katastrophe#chronik-hochwasser-im-ahrtal. Accessed 8 Nov 2021
Weichselgartner J, Brévière E (2011) The 2002 flood disaster in the elbe region, Germany–a lack of context-sensitive knowledge. In: Dowtry RA, Allen BL (eds) Dynamics of disaster–lessons on risk, response, and recovery. Earthscan, London, pp 141–157

Weisenfeld G (2014) Laughing in the face of calamity–visual satire after the Great Kantō Earthquake. In: Juneja M, Schenk GJ (eds) Disaster as image–iconographies and media strategies across Europe and Asia, pp 125–134. Schnell & Steiner, Regensburg

Winckelmann JJ (1767) Anmerkungen über die Geschichte und Kunst des Alterthums. Waltherischen Hof-Buchhandlung, Dresden

Wölfflin H (2015) Priciples of art history. In: The problem of the development of style in early modern art. Getty Trust, Los Angeles

Tamara Schneider is lecturer at Doshisha University in Kyoto, Japan. She holds a Dr. phil. from Kassel University, Germany. Her early research focused on Japonisme, the reception of Japanese art in Europe. At present, Tamara's general interest lies in intercultural relations and nature in art. Currently, she is working on contemporary artists' response to man-made and natural disasters

Chapter 8
The Absence of Gendered Management of Climate Change in China

An Analysis of Adaptation Policies

Ting Wang

Abstract As the world's second-largest economy and the biggest greenhouse gas emitter in total terms, China plays a significant role in addressing climate change and promoting effective action. While gender mainstreaming has been adoptedin the UN system since 1995 and has subsequently been incorporated intodifferent national policies, itis rarely recognizedin China's climate policies. However, enhancing women's participation in climate is an important way of advancing women's empowerment and policy effectiveness.As a young woman with an economics and public administration background and coming from China, my research provides a contentanalysis of national climate change policies in China to examine gender perspectives.I analysehow national policies overlook gender differences in climate change adaptation and frame climate challengeas a scientific and gender-neutral problem. Although gender mainstreaming has been recognized internationally as a strategy in the given policy area, thegender concept is almost absent from China's climate change discourse. Gender is often depoliticized in related plans and strategies. Instead, climate change is primarily associated with economics, technology andsecurity. This notwithstanding, the Chinese government appears to have belatedly recognized the importance of integrating gender into climate issues, as reflected in more recent reports and policy recommendations.

Keywords Gender mainstreaming · Climate change policy · Gender · Policy analysis

8.1 Climate Change, Gender Mainstreaming and Climate Policy in China

Climate change and its devasting consequences–environmental, economic and social–continue to receive heightened attention worldwide. Widespread and rapid changes have been observed in the atmosphere, oceans, cry sphere and biosphere.

T. Wang (✉)
Graduate School of Economics, Kyoto University, Kyoto, Japan
e-mail: wang.ting.84m@st.kyoto-u.ac.jp

M. K. De Santo and S. E. Domptail (eds.), *Degrowth Decolonization and Development*,
https://doi.org/10.1007/978-3-031-25945-6_8

Since 2011, carbon dioxide (CO_2), methane and nitrous oxide concentrations in the atmosphere have continued to rise, reaching annual averages of 410 parts per million (ppm), 1866 parts per billion (ppb) and 332 ppb respectively in 2019 (IPCC 2021).

In the early years of climate science, the primary focus was on identifying the causes of climate change and exploring natural science and technology-based approaches to mitigation.The focus of climate research has evolved to include the assessment of climate change impacts and risks and the potential for adaptation, underpinned by an understanding of the interplay of non-climatic global trends such as biodiversity loss, unsustainable consumption, ecosystem degradation, pandemics and social and economic inequalities (IPCC 2022).

However, climate change impacts and adaptation are not gender-neutral. As gender inequality is one of the most persistent inequalities humanity faces, climate change has had a disproportionately negative impact on women and may further exacerbate existing disparities(Dankelman et al. 2008; Demetriades and Esplen 2008; Preet et al. 2010; Alston and Whittenbury 2013). Despite this, research on the impact of climate change on women remains limited. Several questions resulted in a gender-blind debate such as the division of responsibilities between the developed and developing world, benefits-driven carbon market strategies and the prevailing focus on the causes of climate change from the natural sciences, rather than the social sciences (Fatma Denton 2002).Women have historically been excluded from the economy through the distinction of public/private, exclusion of women from public roles, citizens and power (Klein 1995); and also been excluded from technology through rationality/ irrationality, rationality, logic and objectivity are typically associated with masculinity, but irrationality, emotional, subjective are characterized with women as a strategy in the suppression of women (Oliver 1991).

While gender perspectives in climate change are now beginning to be reflected in research and policies, public doubts remain as to their validity. When presented with a report entitled ‘Women and Climate Change’ by the EU Women's Rights Committee, Marina Yannakoudakis—a British Conservative member of the European Parliament—said: “Global warming is not some male plot to do women down. The climate is the same for males and females, as far as I know. When it rains we all get wet” (Lowther 2012). It is undeniable that climate change is a reality and that both men and women face challenges in sustaining their livelihoods. At the same time, in light of the different responsibilities and social roles of men and women, and the distinct cultural features of diverse societies—the impact of climate change is differentiated depending on exhibits gender-specific differences (Salehi et al. 2015). There is growing evidence that climate change takes different impacts on vulnerability and adaptability between women and men. The extent to which they do so varies between cultures according to their specific social complexities (Eastin 2018; Alston 2014; Trung 2014; Reckien et al 2017). A case study in Gannan Plateau (China) indicate that women have fewer opportunities to receive education but handle nearly all of the household duties as well as farming and herding, which takes more negative impacts on their cognitive ability to climate change, consequently reducing their ability to cope with risks (Wang et al. 2020). In Yunnan province (China), when drought occurred, women have to travel further to collect water for residential and agricultural usage, and their options for pursuing other livelihood actions are restricted as a result, dependence on men is increased (Su et al. 2017).

However, despite this reality, international climate policy has been largely blind to structural gender dimensions. Since the adoption of the Rio Declaration in 1992, little action has been taken to integrate women into the key framework instruments (Rodenberg 2009). In its 44th session, the Committee on the Elimination of Discrimination Against Women (CEDAW) raised concerns about the lack of gender consideration in the United Nations Framework Convention on Climate Change (UNFCCC) and other policies and initiatives, after a study of state parties' reports revealed the gender-differentiated impacts of climate change (CEDAW 2009).While some policies have since endeavoured to narrow the gap, gender perspectives are still not taken into account in a meaningful way; and some market-based solution seven disadvantage women (Gumucio and Rueda 2015). It is thus vital to ensure that climate change interventions and strategies—including policies, grassroots programmes, discussion and research—are gender aware.

China has a crucial role to play in climate change, not only as a potential driver but also as a promoter of effective climate action by considering gender issues in the climate change regime. China is the world's second-largest economy and the biggest greenhouse gas emitter in total volume, accounting for more than 27% of global emissions (Rhodium Group 2020). As a result ofan aggressive industrialization programme, its CO_2 emissions in 2019 were 3.8 times higher than in 1990 and 11.7 times higher than in 1970—figures which illustrate the challenge China faces in becoming carbon neutral (Zhou and Hu 2021, p. 2). While China has made significant progress in addressing climate change challenges recently, gender awareness remains limited and is not adequately reflected in national climate change policies (Zhou and Sun 2020). Equally, there is little gendered analysis of China's climate change policies.

In this chapter, I aim to go some way towards filling this gap. Through a content analysis of China's climate change adaptation policies and plans, the research seeks to determine the extent to which gender perspectives are addressed and framed within them. Based on previous research on gender and climate change, I begin by addressing gendered differences in climate change adaptation and policy limitations. After introducing terminology, I examine whether the dominant climate change policies effectively integrate gender perspectives into climate change issues. I go on to examine existing adaptation policies in China and how these serve to ignore gender roles and relations in climate change. Finally, moving beyond assumptions about women's vulnerability to climate change impacts, I examine proposals for incorporating gender mainstreaming into climate change challenges and policy design.

8.2 Applying a Gender Lens to Climate Change—Climate Change Affects Women and Men Differently

Patriarchy, as a social structure in which men dominate and exploit women, preserves and propagates the belief that men are superior to women, and magnifies gender inequalities and stereotyping (Wood 2019). Conventional patriarchal beliefs as a structure that reward masculine ways of thinking, behaving, and being with socio-economic privileges, construct and reinforce the gendered division of roles. For a long time, masculinism not only is androcentric, but also continues being augmented by Western-centric boasts of superiority, includesin authority, power, scientific rationalism, political acumen, and top-down 'problem solving' (Peterson 2018). Under the patriarchy, women and nature are both devalued and exploited. Women are fundamentally exploited in the capitalist society through unpaid work by which labour is reproduced—the pillar of wealth accumulation and social operation (Federici 2016). Patriarchy makes women more vulnerable to the adverse impacts of climate change and affords men better adaptive capacity, particularly in terms of mobility and flexibility in the employment (Onwutuebe 2019). Compared with men, women have limited ability to manage risks and external shocks, and they are often absent from decision-making and have less power over resources and family finances (Ongoro and Ogara 2012). In the meantime, patriarchal oppression often obscures the differentiated impact of climate change on women and further excludes feminist voices from decision-making, where males dominate in this fieldand determine what is valued and counts.

The vulnerability of different populations to climate hazards is determined by a variety of factors, including the degree of hazard exposure; sensitivity to impacts;coping capacity;gender disparities; and adaptive capacity (Bündnis Entwicklung Hilft 2017). Likewise, individuals are affected differently by climate-induced disasters, due to factors such as gender, age, geographical location, race, social support, and socioeconomic level. Via unpaid work and patriarchy, women confront not only domestic violence, but also institutional burdens that continue to circulate and are inscribed in the economic development (Federici 2016). Many studies have demonstrated that climate change has gender-differentiated impacts (Goh 2012; Ergas and York 2012; Shabib and Khan 2014)—not least because of the different social roles of women and men, which often mean that women are disproportionately negatively affected (Gumucio and Rueda 2015; Alston 2014). Climate change can exacerbate existing gender inequalities through changing livelihoods; it also disproportionatelyimpacts on women's health and well-being and has a detrimental impact on women's opportunities, capacity and work responsibilities (Dankelman et al. 2008).Women displaced by disasters are more vulnerable to specific threats such as sexual and gender-based violence; limited access to healthcare; exploitation; trafficking; fewer opportunities to benefit from technologies and information; and limited resources (De Feyter et al. 2021).Climate change plays the role of a "threat multiplier" in exacerbating inequality. Many studies reveal that climate change has disproportionately detrimental effects on women's health. Because of greater nutritional requirements during menstruation, pregnancy, and childbirth, women often have higher rates of malnutrition and anaemia and are more vulnerable to food crisis caused by climate change (Sorensen et al. 2018). Due to inequalities in cultural norms and basic physiological variables, women and children are more likely to suffer the health risks of household air pollution from inefficient cooking stoves and kerosene lights (Watts et al. 2015). When climate disasters strike, women often suffer more and are less likely to survive; the burden of caring and barriers to accessing resources increased under the negative impacts of climate change (eg.travelling further to collect firewood and water) (World Health Organization 2014).

The evidence further suggests that in agrarian communities, in particular, climate change causes more harm to women and other vulnerable populations because their resources are more constrained (Carr and Thompson 2014). For instance, in the event of a climate-induced crop failure, women and children may suffer greater food insecurity than men because they often have less access to money and labour (Kristjanson et al. 2017).Women may be particularly disadvantaged due to restricted access to resources and justice, limited customary rights, restricted mobility and a lack of voice in decision and policy making (Omolo 2011; Ampaire et al. 2020)When women develop their own adaptive strategies to climate change, they are unable to fully exploit the possibilities they afford due to a lack of personal authority in the household and limited access to resources and knowledge (Djoudi and Brockhaus 2011).

8.2.1 Achieving Gender Mainstreaming in Climate Policy is Challenging

It is thus clear that gender equality should be taken into account as a significant determinant in climate change policy design.Policies should account for these gendered dimensions in terms of bothprocesses and outcomes (Chindarkar 2012). Taking seriously gender and male domination in climate change policies and strategies, identification of gender-targeted actions and equal participation of women are the first essential steps to effective policy-making and implementation (Nyasimi et al. 2018). Taking gender into climate change policies is more than paying attention to the existence of inequality, and is also a significant step to provide light on the key root element that would question patriarchy and change the existing system.

However, despite the importance of incorporating gender issues and the empowerment of women intothe climate change space, the policies to date have given little consideration to this. Gender concerns are barely mentioned in the international policy discourse and women lack adequate representation in climate change decision-making, including in relation to the UNFCCC and the Kyoto Protocol (Terry 2009; Denton 2004). EU climate change policy is likewise gender blindand gender equality is almost absent from the discourse: 'when policy issues intersect, gender disappears'(Alston 2014, p. 16). Constructed via particular socioeconomic and cultural contexts, patriarchy deceived social actors and continues to produce a consensus between women and men, that women's subordinate status and unequal position are natural and normal. Individuals are used to forgetting or being blind to women's vulnerable position in the climate change context. Therefore, even if there is a gender balance in the departments responsible for setting climate policy, greater representation does not guarantee policy change and gender remains conspicuous in its absence from policy documents (Pearse 2017).

There are also numerous barriers to achieving gender mainstreaming in climate policy design and implementation. Although some policies now recognize women's structural vulnerabilities, operational planning still does not address the specific impacts of climate change on women; and while some operational solutions may involve women in the planning stages, few address the specific effects of climate change on women (Shabib and Khan 2014). Although gender awareness may now be improving, it is still largely dismissed as 'women'sissues; also, gender mainstreaming is inconsistent at different levels of government and governments are not yet adequately budgeting for the gender (Ampaire et al. 2020).

Politically problematic dualistic thinking creates androcentric opposition between, like culture/nature, mind/body and rational/irrational, where one side is socially and structurally devalued and women are associated with the devalued terms (Mikkola, Fall 2022 Edition). The dichotomy is used to justify women's oppression and creates men'spositions and power. At the government level, even where gender mainstreaming is mentioned at all, this is unlikely to have a tangible impact on how relevant positions and actor relationships are effectively structured (Acosta et al. 2020). Research already indicates that gender mainstreaming largely ends at the discursive level and gender inequalities are rarely systematically addressed on the ground (Acosta et al. 2019). Considering such inequalities and changing policy frameworkis not achievable through few new measures in the system while the whole status quo is preserved, which needs another true technofix, just as the technologies will save us from climate change. Gender mainstreaming is just a strategy and tool to improve policy regards on gender equality. Challenging existing inequalities necessitates a fundamental reshufflingofvalues and the structures current policymakers shape that organize our society.Here in the face of climate change, addressing gender means addressing not women but thesystematically relative place of women and men in society.

8.2.2 Gender Differences Are Ignored in China's Climate Change Policies?

Over thousands of years, traditional Chinese society was constituted by an essentially masculine value system - an unequal dyadic relationship created a "patriarchal, patrilineal and patrilocal" family system where the male was dominant, and the female was submissive, incapable of education, and without inheritance or property rights (Leung 2003). Until now, these Confucian beliefs of patrilineal descent, patriarchy, and patrilocality are still embedded in the society of Chinese. And this patriarchal system was institutionalized by state policies through supporting the male head of household in resources and services distribution, where men's superiority in authority and wealth was established and women were enforced to subordinate and take responsibility for reproductive labour with little institutional support (Chen 2014).This kind of structural gender inequality is particularly acute in China's rural areas. With the rapid development of urbanization in China, in rural areas, women are often left alone on their husbands' land when men moved to urban areas for work, they take on productive activities in the land, but in the meantime, share the burden of reproduction. However, the development of the world political economy and Chinese policies that fail to include and present rural women who take an important role in both productive and reproductive labour actually exacerbated these deep-rooted gender inequalities (Chen 2014). In the same way, gender inequality persists, restricting the ability to adapt to climate change and reshaping gender relationships through institutions' adaptation interventions (Bhattarai 2020).

Experiences, perceptions, actions and responses to climate change are particularly gendered in China, due to discrepancies in communities (rural versus urban), income, class, geographical conditions and education. Natural catastrophes have gender-differentiated effects, according to research conducted in China's Yunnan Province, with women affected more severely than men (Wang et al. 2020). According to large-scale research on the effects of severe flooding in China, women had a slightly higher risk of developing post-traumatic stress disorder after such events than men (Luliu et al. 2014). On the other hand, women play an essential role in climate change adaptation activities by taking advantage of their acknowledge, abilities and experiences. Chinese men and women have diverse perspectives on climate change, with women being more sensitive to these issues than men (Ginige et al. 2009). Due to gendered roles and duties, women have unparalleled knowledge of their local environment and water circumstances, and their participation is crucial to ensure effective climate change adaptation (Figueiredo and Perkins 2013). Research has confirmed that rural communities with greater involvement of women in forest management activities are more likely to adopt emerging innovations (Zhu et al. 2020). It has further indicated that women often play a vital role in acculturation strategies related to the Clean Development Mechanism forest carbon sequestration programme, and have an advantage in adopting new strategies (Yang et al. 2018).

Although women suffer disproportionately—economically, physically and mentally—as a result of climate change and natural catastrophes, and are increasingly involved in climate change adaptation, there is little gender awareness in Chinese climate change policies. Under the patriarchal system, various statistics reporting established by the Chinese government still do not separate by gender, although gender statistics can help to recognize women's contributions to production and reproduction, as well as mainstream gender equality in policy making (Chen 2016). A comprehensive understanding of women's experiences, as well as policies and actions that are formulated by giving them a legitimate voice, are preconditions of the emancipation of women, but these centralized official policies of gender equality in China, often are rhetorical and ideological rather than substantive (Leung 2003).

Recognizing women's vulnerable considerations and their indispensable contribution in responding to climate change requires not only involving gender awareness in policy but also empowering their potential to participate in the political process and influence policies. However, women's participation and proportion in decision-making and management are much lower than men's in the Chinese government. For example, in the National Committee of Experts on Climate Change, a think tank established in 2007 and consisting of 12 academicians and senior experts from various fields related to climate change to provide advice to the government's decision-making in response to China's climate change, there was no one female member in its first committee. The female members increased a little in the second and third committees, respectively from1 to 4 (proportion of 3 to 10%), still a huge gap (Zhanga et al. 2021, p. 102). When women are increasingly involved in climate change activities, corresponding policies supporting their role have yet to be developed in China.

8.3 Gender Mainstreaming in Climate Change Adaptation

8.3.1 What is Gender Mainstreaming?

Gender mainstreaming has been advocated through the development of plurilateral governance, such as the United Nations and the European Union, and through a variety of transnational discourses on human rights. In 1997, the UN Economic and Social Council defined 'gender mainstreaming' as follows.

Mainstreaming a gender perspective is the process of assessing the implications for women and men of any planned action, including legislation, policies, or programs in all areas and at all levels. It is a strategy for making women's as well as men's concerns and experiences an integral dimension of the design, implementation, monitoring, and evaluation of policies and programs in all political, and economic societal spheres so that women and men benefit equally and inequality is not perpetuated. The ultimate goal is to achieve gender equality (UN 1997).

Nearly 30 years after the 1995 adoption of the Beijing Platform for Action—which aimed to achieve 'gender equality and the empowerment of women', and prioritized gender mainstreaming—its commitments have not been translated into action (Caglar

2013). As a strategy and tool to promote political action, there are many reasons resulting in gender mainstreaming beingoverlooked and suppressed, including its own flaws and also systemic problems. First, a lack of conceptual clarity resulted in gender mainstreaming uncertainty and ambiguity (Daly 2005). Furthermore, political players have not clearly articulated what gender equality is, and in which ways to achieve gender equality, resulting in divergent perceptions of gender mainstreaming in policies (Caglar 2013). Even within those organizations, which are responsible for supporting gender-mainstreaming policies, there is no agreed definition of 'equality' (Meier and Celis, 2011). There are also political and institutional barriers to gender mainstreaming.Gender mainstreaming aims to alter institutional frameworks, policy tools, and precedences from a gender equality perspective, but it has not naturally brought about socialization processes and is being opposed by powerful institutional and administrative forces (Caglar 2013). It is sometimes constrained by accusations of elitism and individualism when conceptualized as a bureaucratic policy tool towards equal opportunities (Squires, 2005). Although EU actors frequently pay lip service to the importance of gender mainstreaming, it is notably absent from primary policy and is often treated as a procedural rather than substantive matter (Meier and Celis 2011). Achieving gender mainstreaming across overlapping policy domains is more difficult still. Meier and Celis (2011) suggest that gender mainstreaming should be considered not only in relation to climate change and other significant issues but also in relation to other cross-cutting matters. In addition, due to institutional reluctance and lack of commitment, the gender mainstreaming approach still suffers from a gap between the policy design and implementation (Parpart 2014).

On the other hand, when gender mainstreaming is introduced, it does not root in analysing gender inequality as a structural problem, rather selectively utilizing as a tool and procedure of its basic techniques, motivated by the wish to improve equality architecture in the light of "fashions" in policy-making (Daly 2005). While gender mainstreaming is indeed an important strategy to partly alleviate women's plight within existing patriarchal capitalist power structures, it does little to question and transform the very grounds of that structure.

Beier (2018) defined a process called "double enclosure", through aiming to transfer women into the paid workforce, reproductive activities are devalued, which consequently creates a precondition for further commodification of women's care and domestic tasks and domestic work. She mentioned that this kind of reducing social reproduction tasks by increasing efficiency rather than redistributing tasks between women and men, making some achieving women's 'economic empowerment' policies actually only contribute to the goal of recognition, but further aggravates inequality and, "ironically it appropriates of feminist theory and practice, which identified 'wages for housework' and the economic role of social reproduction (Beier 2018)." Gender mainstreaming is an example of this kind of women's economic "empowerment" political tool, we should maintain vigilance to this appropriation, which may likely exacerbate inequalities. The change brought by gender mainstreaming in political agency and systematically achieving gender equality is limited.

However, such strategies are urgently required because without them, women will remain disenfranchised and excluded from public economic activities and political life (Porter and Sweetman 2005).The socially assigned roles, responsibilities and affiliations of women and men are ignored in a gender-blind policy design approach. By contrast, gender-sensitive strategies acknowledge the different roles and positions of men and women in society, recognizing that they have different demands and interests (Gumucio and Rueda 2015). Gender mainstreaming is an institutional tool for achieving gender equality by rejecting andocentric notions and traditional female roles (Squires 2005). The role of governments in addressing women's marginalization and exclusion through economic, political and social reforms cannot be overlooked; and a gender-sensitive approach in policymaking on climate change is thus essential.

8.4 Importance of Climate Change Adaption

Climate change is predicted to intensify in the coming decades and weather patterns will become more variable as a result. Extreme weather events—such as tropical storms, droughts, floods and cyclones—are expected to become more frequent and possibly more severe. While China's economy has expanded rapidly in recent decades, the country has just 7 per cent of the world's arable land available and 22 per cent of the world's population to feed, making it particularly vulnerable to these effects of climate change. This is exacerbated by its geographic position, high population density, economic structures and reliance on climate-sensitive industries for livelihoods (Piao et al. 2010). Under socioeconomic pathways, anticipated drought losses in China at the 1.5 °C warming level will respectively increase tenfold and almost threefold comparing the reference period 1986–2005, and the span 2006–2015 (Su et al. 2018).Facilitating adaptation and reducing vulnerability are thus vital aspects of climate change policy—especially in China.

The term 'adaptation' refers to adjustments to a system's characteristics and behaviour that strengthen its ability to cope with existing or expected external pressures (Brooks 2003). Increasing the ability to adapt to climate change and reducing social vulnerability to its impact will help to minimize the risk of climate hazards. However, as a developing country, China has to deal with its own double adaptation challenges: the need to consider economic development and the pressure of achieving it through a sustainable way. Although policies for coping with climate change adaptation are experiencing a mainstream trend in China, there is still a mismatch between the goals of adaptation and allocation of resources, and implementation challenges regard to evaluation and supervision (He 2017).

From a gender perspective, women face greater challenges in adapting than men, due to gendered responsibilities assigned through the conventions, rules and practices of patriarchal cultures, such as reproduction and unpaid family work division. Even integrating women into the paid workforce and the commodification of care and domestic duties, if only focusing on increasing efficiency, rather than redistributing tasks, will increase gender inequality further, particularly between women along the lines of ethnicity/race, south/north, and classes (Beier 2018). The fact is that the escalation and deepening of the market-centred framework, international competition and financial capital barricades impoverished women from obtaining their societal, legal, and economic rights (Tandon 2015).

In recent decades, men who previously farmed have been migrating from rural areas to find work in China's cities. Consequently, the number of women farmers in China has increased; but they often find it difficult to earn enough income due to their limited land and resources and increased household burden (Zhao et al. 2015). A case study in China Gannan Plateau found that in Tibetan households, women are responsible for nearly everything, from farming and herding to household duties (Wang et al. 2020). Failure to recognize that disadvantaged groups face greater challenges to adaptation may lead to discriminatory outcomes, such as fewer adaptation resources and solutions, rendering these groups even more vulnerable to climate change (Mersha and Van Laerhoven 2016). Under the background that women actually play more important roles in climate change adaptation activities, governments should take responsibility for policy design intervention to dismantle gendered barriers to adaptation.

8.4.1 A Content Analysis of Policy Documents

This research aims to examine whether gender-based approaches and viewpoints are effectively considered, and the extent to which gender mainstreaming has been achieved, in China's climate policies. In addition to a secondary literature review, the research is based on an examination of selected policy, planning and regulatory documents from China.

Inspired by Shabib and Khan (2014) research on gender-sensitive adaptation policymaking in Bangladesh, the research adopts the following methodology to achieve its goal. The starting point was to gather existing literature and documents on China's climate change adaptation policy in order to analyse the extent to which gender perspectives are discussed or mentioned. I compiled a list of all relevant policies based on the annual *China's Policies and Actions for Addressing Climate Change* reports published from 2008 to 2020. The climate change adaptation section of these reports covers annual policy documents that have been published in six fields: *agriculture; water resources; forests and other ecosystems; coastal zones and coastal ecosystems; human health; and disaster prevention and mitigation, risk management and early warning of climate disasters.* I then assessed the extent to which the terminology of gender is reflected in China's climate change adaptation policies, to highlight policy limitations relating to gender. The information drawn from national government policies was supplemented with data from relevant journals, articles, reports, public documents, websites, records and other sources.

I then developed some guidelines (Table 8.1) to analyse all of the documents described above through a gender lens. The guidelines were based on accepted gender mainstreaming principles and tailored to the needs of this particular study.

Table 8.1 Guidelines and criteria for gender-sensitive climate change adaptation policy

Guidelines	Criteria
1. Gender difference	Whether climate change policies/plans acknowledge gendered difference in terms of climate change impacts and vulnerability; and whether gender analysis on adaptation has been conducted
2. Gender norms	Whether vulnerable groups are defined and targeted by involving gender factors, and whether gender concepts are reflected in policy documents
3. Responsibility	Whether all staff share responsibilities for adaptative strategy implementation, with support from gender specialists
4. Gender equality	Whether the design of climate change policies takes account of actions targeting gender equality and responds to the needs of vulnerable groups
5. Participation and empowerment	Whether policies encourage and broaden women's participation at all levels of climate change activities and decision making; and whether they recognize women's strength and capacity, and support women's empowerment through capacity building and training
6. Gender training and implementation	Staff understanding, sensitization and implementation of policies: whether climate change adaptation strategies and actions are identified based on local knowledge, resources, accessibility, availability and vulnerable groups
7. Monitoring and evaluation	Accountability for outcomes: whether the monitoring system identifies feedback and evaluation outcomes of actions; whether activities successfully benefit gender equality
8. Budgets	Whether policies allocate human and financial resources to translate gender mainstreaming concepts into practice

Source Policy resources for content analysis

Guidelines for the assessment of policies

According to the Report of the UN Economic and Social Council for 1997(UN 1997), the principles for gender mainstreaming can be summarized as follows:

- Gender differences: 'An assumption of gender-neutrality should not be made.'
- Responsibility: 'Accountability for outcomes needs to be monitored constantly.'
- Participation: Women's participation at all levels of decision-making should be broadened.
- Institutionalization: Gender mainstreaming should be institutionalized through concrete steps, mechanics and processes.
- Irreplaceability: Gender mainstreaming cannot replace women-specific policy programs, positive legislation, gender units or focal points.

- Commitment: Clear political will and adequate human and financial resources are needed to translate gender mainstreaming concepts into practice.

It is critical to adopt the terminology of gender equality and gender mainstreaming in policymaking. According to Caroline Moser and Annalise Moser's (Moser and Moser 2005) summary of gender mainstreaming, such policies should include the following fundamental components: A combination of gender mainstreaming strategies with gender equality actions; Gender analysis; Responsibility sharing and involvement of gender specialists; Gender training; Support of women's empowerment and decision-making; Monitoring and evaluation; Cooperation with other organizations; Budgets;Knowledge resources.

Based on these components, and inspired by the analysis principles and criteria for gender inclusion in Africa's climate change policies (Nyasimi et al. 2018), I compiled a set of gender mainstreaming guidelines and criteria (Table 8.1) to analyse whether gender has been integrated into China's climate change policies.

Table 8.2 presents national adaptation policies that can be analysed according to 6 categories of agriculture, forestry and the ecosystem, water resources, public health, marine resources and disaster prevention and mitigation systems, which include regulations, planning and strategies.

8.4.2 Analysis: China's Climate Change Adaption Policies

A review and analysis of gender mainstreaming in China's climate policies begin with a consideration of whether gender equality and gender mainstreaming terminology have been incorporated into the respective policy documents. China is a key player in international climate issues and has significant influence and status in global climate negotiations (Heggelund 2007). Although China's leadership has become increasingly aware of climate change issues, this has not yet surpassed economic development as a policy priority (Lewis 2007). Published in 2007,*China's National Climate Change Programme* established six guiding principles in relation to climate change (NDRC 2007): to address climate change within the framework of sustainable development; to follow the UNFCCC principle of 'common but differentiated responsibilities'; to place equal emphasis on both mitigation and adaptation; to integrate climate change policy with other interrelated policies; to rely on the advancement and innovation of science and technology; and to participate in international cooperation actively and extensively.

The following specific objectives were emphasized in the Program.

Control of greenhouse gas emissions: reducing energy consumption by 20 per cent per unit of gross domestic product; increasing the proportion of renewable energy to 10 per cent by 2010; and increasing the forest coverage rate to 20 per cent.

Capacity to adapt to climate change: increasing grassland and the utilization coefficient of agricultural irrigation water; ensuring that nature reserve areas account for 16 per cent of the total territory; and controlling decertified lands.

Table 8.2 Selected national adaptation policies in China

Categories	Year	Regulations/planning/strategies
Agriculture	2012	Opinions on promoting water-efficient agricultural development
	2012	State Council's outline of national agricultural water conservation (2012–2020)
	2012	Opinions on implementing the strictest water resources management system
	2014	Measures to implement the national drought control plan
	2014	Notice on promoting dryland agriculture technology
	2016	National agricultural modernization plan (2016–2020)
	2017	Preparatory plan for disaster mitigation and relief in a science-based response to El Niño to ensure a bumper harvest
	2017	Measures for the management of subsidy funds for the protection of agricultural resources and ecology
	2019	National agricultural modernization plan (2016–2020)
Forestry and ecosystem	2012	Action points for China's forestry departments in response to climate change during the 12th five–year plan period (2011–2015)
	2016	Action programme on forestry adaptation to climate change (2016–2020)
	2016	National forest fire prevention plan (2016–2025)
	2016	Rehabilitation plan for farmland, grassland, rivers and lakes (2016–2030)
	2017	Five-year plan for grassland protection, construction and utilization
	2020	National master plan for key ecosystem protection and restoration projects (2021–2035)
Water resources	2011	National urban drinking water safety protection plan (2011–2020)
	2011	National rural drinking water safety project for the 12th five-year plan period (2011–2015)
	2012	Opinions on implementing the strictest water resources management system

(continued)

Table 8.2 (continued)

Categories	Year	Regulations/planning/strategies
	2012	Guidance on strengthening the supervision and monitoring of drinking water quality
	2012	2012 National drinking water supervision and monitoring work plan
	2014	Notice on further strengthening the supervision and monitoring of drinking water quality
	2019	Guiding opinions on defining and guaranteeing the ecological flow of rivers and lakes
	2019	Overall plan for smart water conservancy
	2019	Three-year action plan for improving the level of water conservancy network information (2019–2021)
	2019	National water conservation action plan
	2020	Guiding opinions on defining and guaranteeing the ecological flow of rivers and lakes
Public health	2007	National environment and health action plan (2007–2015)
	2011	National urban drinking water safety protection plan (2011–2020)
	2013	Urgent notice on offering emergency public health services in the wake of natural disasters
Marine resources	2011	State council-approved national plan for marine functional zones (2011–2020)
	2012	Marine medium and long-term plan in response to climate change (2011–2020)
	2016	National scientific and technological actions on climate change during the 12th five-year plan period (marine area)
	2017	Development plan of national maritime economy during the 13th five-year plan period
	2020	Special action plan for mangrove protection and restoration (2020–2025)
Disaster prevention and mitigation systems	2009	National meteorological disaster defence plan (2009–2020)
	2011	National disaster prevention and mitigation plan (2011–2015)

(continued)

Table 8.2 (continued)

Categories	Year	Regulations/planning/strategies
	2014	National commission for disaster reduction and national bureau of statistics
	2015	Measures for the management of national emergency early warning information release system
	2016	National comprehensive disaster prevention and mitigation plan (2016–2020)
	2016	Action plan for urban adaptation to climate change
	2017	Measures for the administration of subsidy funds of the central finance for the relief of agricultural production disasters and the management of exceptionally serious floods and droughts
	2017	Work plan for thoroughly implementing the opinions of the central committee of CPC and the state council on promoting the reform of disaster prevention and mitigation systems and mechanisms
	2018	Opinions on strengthening disaster prevention, mitigation and relief from the meteorological perspective
	2019	Opinions on promoting the reform of systems and mechanisms for disaster prevention
	2020	Implementation plan for resolving weak links in flood control during the 14th five-year plan period
	2020	Technical requirements for geological hazard risk investigation and evaluation (Trial)
	2020	Technical guide for geological disaster special group combined monitoring and early warning (Trial)

Enhancement of research and development; and Public awareness and management (NDRC 2007).

Climate hazards are particularly dangerous for vulnerable regions and individuals. For example, the human mortality caused by flooding, droughts and storms in highly vulnerable areas was 15 times greater than that in less vulnerable areas between 2010 and 2020 (IPCC 2022, p. 14). While ostensibly adhering to the principle of giving equal weight to both mitigation and adaptation, seen from the specific objectivesof the Program (NDRC 2007) in practice, China still prioritizes mitigation over adaptation—particularly through technological advancements and large-scale infrastructure improvements.

China has published a report entitled *China's Policies and Actions for Addressing Climate Change* every year since 2007. A review of these annual reports makes clear that China's climate change adaptation policies have been refined over time, particularly between 2007 and 2020, with a focus on six sectors below:

- agriculture;
- forestry and the ecosystem;
- water resources;
- marine resources;
- public health; and
- disaster prevention and mitigation systems.

My research analyses the extent to of gender equality and gender mainstreaming terminology have been incorporated into the respective adaptation policies (Table 8.2) in six categories respectively.

Agriculture and water resources

China's national climate change programme identifies agriculture and water resources as two critical sectors for adaptation, since water resources are the most important factor that influences agricultural development. China has long highlighted the significance of expanding regional water storage and enhancing water supply and management infrastructure as adaptation strategies to ensure agricultural security and protect ecosystems from the effects of climate change, such as the South-to-North Water Diversion Project, a hydraulic project being planned to help with water resource distribution; flood management; and drought relief in the north (Piao et al. 2010).

A review of China's agricultural policies reveals that adaptation strategies also primarily focus on infrastructure and technological development, with the following aims: strengthening the response to negative impacts on agricultural production caused by climate change-induced flooding and drought; preventing crop diseases and insect infestation; creating new large-scale irrigated lands and renovating existing facilities; promoting water-saving agriculture, drought-resistant dry agriculture and moisture-conserving adaptation techniques; and improving the quality of culverts.

It can be seen that there is no any mention of "gender training" in promoting adaptation techniques or "gender norms and differences" in terms of the negative impacts of climate change. Actually, the human factor is largely ignored in these adaptation policies, which lack participatory sociological concepts and gender perspectives. There is no consideration of either the vulnerability or empowerment of women.

My review of China's adaptation policies relating to water resources also reveals that policies pay more attention to developing large-scale infrastructure; gender, even people, have not been taken into account. Under the central government-led, a top-down climate governance, the main initiatives and content presented in China's water resource strategies for climate adaptation involve developing projects for flood control, drought relief and disaster mitigation systems; implementing a nationwide water resources management system; establishing hydrological and water resources monitoring systems; designing nationwide plans for water resources and river basins; and pursuing water diversion projects such as the South-to-North Water-Diversion Project. All these big projects and strategies met the characteristics of China's "authoritarian environmentalism", in which policies and programs are launched by limited "elites", to resolve climate problems intensively and efficiently with a strong state (Gilley 2012). Centralising resources and focusing on technological advancement in China's climate change adaptation is an expression of masculinism in practice. Masculinism continues to be presented and augmented here by emphasizing a kind of authoritative, scientific-rational, efficient, representative top-down, and "knowledge claims," climate change adaptation solutions. Under this system and government environment, gender-specific mitigation solutions and women's empowerment in public activities relating to water management are not addressed.

When it comes to setting water management policies and designing climate change adaptation methods, gender mainstreaming has received little attention in China. When adaptation responses are not matched to women's needs and interests, women's positive participation in water management is precluded (Su et al. 2017). In the policies relating to water management and planning described above, there is no gendered evaluation of climate change impacts, which hinders the ability to adapt to water-related crises caused by climate change. Recognizing gendered differences and involving women in water resource management would enable women to serve as active agents in adapting to climate change, but these gender perspectives are not acknowledged in these documents. The research confirms that government policy tackles neither gender concerns relating to climate-induced water issues nor gender-sensitive adaptation strategies.

Forestry and the ecosystem

The research reveals that China has adopted broad adaptation activities and objectives in relation to forestry and ecosystem policy. The primary consideration here is how to optimize forest structure and functions, including by changing animal husbandry methods; promoting the conversion of farmland to forests and grasslands; awarding subsidies; preventing natural disasters; controlling harmful pests, and promoting ecological protection.

With regard to gender, I found that policies focus on how to optimize the forest structure in view of climate change, rather than on how people interact with forests and how this relationship will change due to global warming. The *Action Plan for Forestry to Adapt to Climate Change (2016–2020)* emphasizes the importance of the government as the major innovator and guide to lead others in adapting to climate change:"Adaptation is the focus of forestry to cope with climate change and should cultivate specialized personnel in adaptation. Efforts should be made to publicize climate change-related knowledge and improve policy understanding. The government should actively carry out trials and demonstrations of forestry adaptation to climate change (p5)." In China, environmental policies are produced by following the model called "'Authoritarian environmentalism"—a non-participatory approach where limited 'participation' by scientific and technocratic elites, termed as 'ecoelites' steer state policy and implementation (Gilley 2012). Same to the adaptation policies here, government-led "ecoelites" are in charge of formulating policies, and strategies to take a rapid response to the comprehensive environmental challenges. In terms of "cultivate specialized personnel," I did not find any words in the document related to "gender training or encourage women's participation", which are two important criteria in Table 8.1f or analysing gender sensitivity in policy.

In addition, people, and of course not their gender, are not explicitly mentioned by the policy texts, gender mainstreaming is thus still a long way off. And this also proves the characteristic of the patriarchal system—focusing on technology and technical solutions. Adaptation policies and strategies which ignore existing gender inequities in forest management, seek solutions consistent with consolidating masculinity, are likely to constrain women's potential, resulting a "waste" of women's ability in forest management and climate change adaptation, and hamper forest resilience.

Public health

Climate change has direct and indirect health consequences for individuals, particularly those who are already vulnerable and at risk. Heat stress from rising temperatures causes heart attacks, strokes, respiratory failure and heatstroke (ESCAP 2018). According to ESCAP, heat wave mortality and morbidity are increasing, with urban residents in informal settlements, socioeconomically disadvantaged groups and minority ethnic groups particularly vulnerable to heat risks (ESCAP 2018).

Especially in rural areas, climate disasters, poverty and hazard vulnerability create a vicious circle, women have enhanced vulnerability because of their physical constraints (pregnancy, delivery, nursing care, etc.,) and economic, and social inequality (Kesavan and Swaminathan 2006). Climate change events may also disrupt the usual timing of menarche, resulting in disruptions that could potentially increase the illness burden for women (Canelón and Boland 2020). However, gender is either undervalued or non-existent in both research and policy documents on climate change and health. Even the World Health Organization materials lack a gender lens in their methodology and climate policy suggestions (Preet et al. 2010).

Equally, in China, national policies on health and adaptation to climate change lack any gender analysis and specific action is non-existent. *National Environment and Health Action Plan (2007–2015)* mentions establishing a monitoring network for extreme climate events and health issues: by using the national meteorological system to establish a monitoring network and strengthen forecasting capacity for extreme weather, such as heat waves, floods, droughts, storms, sandstorms and cold waves, etc. Besides, there are policies that emphasize national supervision of drinking water quality and monitoring of disease prevention caused by climate change. All policy documents consider health issues from a macro and general perspective, and gender-specific health consequences are not acknowledged in related policy design.

Positively, the newly published *National Climate Change Adaptation Strategy 2035*[1] does acknowledge the need to pay more attention to especially vulnerable groups:

> based on the results of the health risk assessment of climate change, to take into full account the climate characteristics of each region and the health risk exposure level of vulnerable populations, conduct the assessments of the adaptation capacity of medical and health systems and key vulnerable populations to climate change, and formulate plans to improve adaptation capacity.

The strategy mentions formulating plans and guidelines for assessing climate change health risks and establishing a comprehensive assessment mechanism for effectiveclarification of who are vulnerable populations. However, gender has not been a criteria forrecognising a "vulnerable population", admitting "gender difference" and presenting "gender norms" are important parts of gender-sensitive policies (Table 8.1). Adaptation policies on climate change in China have not included differential effects of climate change on the health of women and men, and gender still is not considered in strategic health policy planning in China.

Marine resources

Climate change and global warming have a significant impact on marine activities. Women play a vital role in preserving food security in coastal areas affected by extreme weather, as when fish catches are insufficient to support the family, women are often responsible for seeking alternative food sources. They are thus often at risk of shouldering a bigger provisioning load (Savo et al. 2017).

China's government only published a few policy documents related to marine adaptation to climate change. Once again, climate change adaptation policies in China have not integrated gender awareness into marine resource policy design. Instead, all actions and plans are concentrated on macro-targets and infrastructure construction related to preventing and penalizing marine environmental pollution, including strengthening coastal ecological restoration and vegetation protection; building coastal forest shelterbelts and implementing tide prevention measures; and enhancing coastal zones and ecosystems. Although men have historically done most of the fishing, women have also played a key role in the catching and gathering of marine resources. From a global perspective, women mainly undertake ' gleaning'—fishing from the shore or reefs, and collecting shellfish and other invertebrates—which supports subsistence and regional market needs and provides significant nutrition (Schwerdtner Máñez and Poulsen 2016; De la Torre-Castro et al. 2017). Even though men and women mostly fish in different locations and have different responsibilities, the reviewed policies in China are nonetheless gender-blind.

Considering women play important role in marine activities, good adaptation policies and measures should involve women and share their experiences and advice.

---

[1] The State Council of the People's Republic of China (2022). National Climate Change Adaptation Strategy 2035.

http://www.gov.cn/zhengce/zhengceku/2022-06/14/5695555/files/9ce4e0a942ff4000a8a68b84b2fd791b.pdf.

China has abundant marine resource demand and also has its specific cultural, socioeconomic, and political conditions, when facing the challenge of climate change, specific research and empirical data are needed in policy design for including gender mainstreaming in adaptation policies. On the contrary, related data and empirical research are very limited in China, realizing and understanding what happened on the ground and in reality is important, the lack of this kind of research and existing "gender-neutral" policies make women's roles more invisible. In addition, avoiding some outdated generalizations, and formulating policy based on own national conditions and reality is especially important when including gender perspective in policymaking. For example, realizing the importance of traditional knowledge in adaptation activities and overcoming biases in policies that prioritize "western measures of gender equity (e.g., number of women in government) and scientific knowledge, over traditional knowledge (Mcleod et al. 2018)". Also, avoiding out dated opinion is necessary, such as presuming women are more environmentally conscious and so support more 'selfless' conservation attitudes, but the fact from the sample is that they felt less connected to the sea environment (Graziano et al. 2018). Now, the related policies are still gender-blind, and respectively related research gaps still exist in China.

Disaster prevention and mitigation systems

It is clear that gender is an important factor that should be considered in disaster adaptation policies. However, only in China's *National Commission for Disaster Reduction and National Bureau of Statistics*, published in 2014, are women included alongside the elderly, children and 'SANWU' people (ie, those without identification papers, a normal residence permit or a source of income) under the heading 'people requiring life assistance' as vulnerable groups (UNWOMEN 2016). China's current climate change policies instead emphasize monitoring and early warning mechanisms for natural disasters and extreme weather and climate events; technologies to prevent and mitigate the effects of agricultural disasters; agricultural subsidy policies; disaster relief measures; public education on disasters; and a forecast service for impending crises.

Concerns about the impact of disasters on vulnerable populations have been highlighted in policies, laws, planning and regulatory documents. In this regard, China's adaptation plans recognize the need to prioritize adaptation efforts in sensitive areas and among vulnerable populations, and to strengthen special services for vulnerable groups in case of extreme weather events. However, gender is not yet reflected in Chinese policy as an important measure of vulnerability. An awareness that certain groups are more vulnerable to the negative impacts of the climate crisis does not equate to the inclusion of gender awareness in climate issues. None of these policies includes terminology of gender; instead, the neutral term 'person' is primarily used across all policies, and there are no provisions that specifically address gendered impacts or gender-sensitive adaptation. This analysis confirms that the different impacts of climate change on women and men are not reflected in climate change policies, and that gender mainstreaming thus remains a long way off.

8.5 Conclusions and Proposals

Women often experience more negative impacts from climate change due to a variety of social and physiological factors. It is thus vital to recognize that gender equality is a cross-cutting issue that influences a variety of the UN Sustainable Development Goals (SDGs), especially climate action. Achieving gender mainstreaming in climate change policies and empowerment of women would help to improve overall sustainable development. Through a content analysis of climate change adaptation policies and plans in China, this research has analysed whether gender perspectives are integrated into these policies and plans, and the extent to which gender mainstreaming has been achieved in them. The main findings and conclusions are as follows:

Climate change events are predicted to become both more frequent and more variable in the future, leading to a need for social shifts and adaptation strategies. Effective climate change adaptation policies should recognize that men and women experience different climate change impacts due to the different roles they play in society. Viewing China's climate change adaptation policies through a gender lens, it is clear that many of them remain resolutely gender blind. The word 'woman' barely appears in any of these policies. The reality is thus that China has neither included gender perspectives in national climate change policy nor addressed the vulnerabilities of women or the empowerment of women in climate change actions.

On the contrary, the prevailing worldview/perspective in which the policies are written remains resolutely patriarchal. The adaptation policies have a limited participatory and often are considered from an energy and scientific-technical perspective. The government claims a kind of authoritative, scientific-rational, top-down solution to intensively respond to climate challenges. Related discussions, strategies and plans concentrate on the technocratic, general guidance and regulatory, rather than society, and people.

Gender-sensitive operational strategies to address climate change are also absent. At present, there are very limited statistics on gender and climate change in China, whether in government statistics and research data or academic research papers and monographs. This needs to be urgently addressed. The research reveals that adaptation policies are mainly focused on technology and engineering strategies; and that although some policies and strategies emphasize support for vulnerable groups, this does not extend to women.

Gender issues in China's climate change regimes have not been prioritized. While gender inequality leads to social injustices and wastes potential human resources, especially at times of crisis, achieving gender mainstreaming in climate change politics still appears to be something of a luxury, especially in China.

Although the framework for understanding climate-related risks and vulnerabilities initially concentrated on the geography and technicalities of climate change, more recent approaches are exploring other ways of interpreting vulnerability within and across population groups (ESCAP 2018). Scholars and governments are paying more attention to gender issues and how they relate to climate change. Although there is no mention of gender in its climate change adaptation policies, China has committed both domestically and internationally to achieving both gender equality and sustainability (CCICED 2021c), including through: the Convention on the Elimination of All Forms of Discrimination Against Women; the United Nations Conference on Environment and Development, AGENDA 21; the Beijing Declaration and Platform for Action (Beijing, 1995); the Millennium Development Goals; the Outline for the Development of Chinese Women (2011–2020); and the 2030 Agenda for Sustainable Development and the Sustainable Development Goals.

Both the Chinese government and Chinese academics have begun to pay attention to the impact of climate change on women. In 2011, the government established the Program for the Development of Chinese Women (2011–2020), which recognized the rights of women right to development in the fields of economy, environment and law, among others; and which explicitly stated that women would be given more development opportunities in terms of 'participation in decision-making and management' (Zhou and Sun 2020).

Although specific strategies and measures in this regard have not yet been included in official documents, gender awareness is gradually being integrated into climate change policy design. Women's role as active agents in addressing climate change is recognized in the *2021 Special Policy Report on Global Climate Governance and China's Role* (CCICED 2021a).The report highlights the importance of gender equality in combating climate change and the link between gender equality and sustainable development—both of which should be further recognized in Chinese policy. In order to mainstream awareness and action towards gender equality in the implementation of climate policy, another report recommends that the government consider gender issues when updating its nationally determined contribution under the Paris Agreement (CCICED 2021a). The China Council for International Cooperation on Environment and Development's (CCICED) *Annual Policy Report 2020: Green Consensus and High-Quality Development* emphasizes that gender equality and empowerment of women will have beneficial effects in achieving the other SDGs—particularly those relating to energy access and climate issues.

It also presents several corresponding policy recommendations, including: systematically collecting information and consulting with communities at the local level, and providing more 'life quality-improving choices for clean energy access' by strengthening women's involvement; introducing innovative financing schemes that encourage women's participation in the evaluation, management, administration and financial benefits of distributed renewable energy and energy storage, thus resulting in complementary income; and raising awareness of the impacts of climate change on women (CCICED 2021b).

Likewise, the CCICED's 2021 Policy Recommendations suggest that gender equality should be factored into the green transition in China, especially in relation to employment and economic development in coal-reliant areas; and that increased support should be provided to women farmers—for example, in the form of knowledge and skills training in agricultural innovation and technological application (CCICED 2021b). Although these are merely policy recommendation reports, they indicate that a new awareness of gender issues in climate change is developing in China.

Including gender awareness in the adaptation, policy design is the first step to achieving gender mainstreaming, and these plans show the possibility that policies consider gendered differences in climate change adaptation, which will help to reduce the vulnerability of women. Some targeted knowledge and skills training for women, on the practical level, will empower women in economic participation. However, including women in decision-making has not been seen. More importantly, the underlying causes of the marginalization of women in adaptation strategies and the fundamental inequities that create women's specific vulnerability are not discussed. Without changing the existing gender division, and not seeking to address existing gender roles and responsibilities, is hard to lead to challenging this patriarchal system.

In sum, while the policy documents issued up to 2020 analysed in this document study show no consideration of gender issues, recent initiatives in China seem more promising. However, whether they will have any tangible effect remains to be seen. It is clear, however, that China's climate change policy should become more gender-sensitive, given the different ways in which women are impacted by climate change in China. Women's roles in climate change should thus be recognized and strengthened, not only through improving understanding of gender-differential vulnerabilities to the impacts of climate change but also through empowering women and facilitating their involvement in adaptive strategies and planning, which can help China to achieve gender mainstreaming in climate change policies. To this end, the government should conduct gender and vulnerability analysis, and establish sex-disaggregated baselines, indicators and targets (CCICED 2021a).

If gender sensitivity were incorporated into China's climate change policies, this would advance gender mainstreaming and contribute to the formulation of more effective and comprehensive climate change policies in China and beyond. However, what is more, introducing and achieving gender mainstreaming in policymaking is not the final purpose, as a tool and procedure by utilizing its basic techniques to alleviate women's plight, gender mainstreaming cannot fundamentally question and transform the grounds of this patriarchal power structure. To achieve gender equality and improve policies, mechanisms also should be established to improve women's political participation and to obtain equal voices in processes of effective responses to climate change, including measures aimed at environmental conservation, disaster risk reduction, natural resource management, and sustainability consumption. gender mainstreaming is a tool not only to alleviate gender inequality but also to respond to climate change challenges better. How best to harness and enhance women's abilities and powers in addressing climate change issues is a challenge, but also an opportunity for sustainable development.

Acknowledgements I am grateful to all of those with whom I have had the pleasure to work. I would especially like to thank Dr Sven Rudolph. As my mentor, he taught me a great deal about research and always gave me a lot of encouragement during my doctoral study. I am grateful to Prof. Turo Morotomi, my supervisor at Kyoto university, thanks for his help with my research. I am especially indebted to constructive comments and suggestions from Dr Stephanie Eileen Domptail and Laura Zuber, their help led to a significantly improved paper. I thank Dr Tamara Schneider for her suggestions for my paper. I also gratefully acknowledge Otsuka Toshimi Scholarship Foundation's support for my doctoral study and research. Any views expressed or remaining errors are solely the responsibility of the authors.

References

Acosta M et al (2019) Discursive translations of gender mainstreaming norms: the case of agricultural and climate change policies in Uganda'. Women's Stud Int Forum 74:9–19. https://doi.org/10.1016/j.wsif.2019.02.010

Acosta M et al (2020) The power of narratives: explaining inaction on gender mainstreaming in Uganda's climate change policy. Develop Policy Rev 38(5):555–574. https://doi.org/10.1111/dpr.12458

Alston M (2014) Gender mainstreaming and EU climate change policy. Women's Stud Int Forum 47(PB):287–294. https://doi.org/10.1016/j.wsif.2013.01.016

Alston M, Whittenbury K (2013) Gendering climate knowledge for justice: catalyzing a new research agenda. In: Research, action and policy: addressing the gendered impacts of climate change, pp 1–281. https://doi.org/10.1007/978-94-007-5518-5

Ampaire EL et al (2020) Gender in climate change, agriculture, and natural resource policies: insights from East Africa. Clim Change 158(1):43–60. https://doi.org/10.1007/s10584-019-02447-0

Beier F (2018) Marxist perspectives on the global enclosures of social reproduction. TripleC 16(2):546–561. https://doi.org/10.31269/triplec.v16i2.980

Bhattarai B (2020) How do gender relations shape a community's ability to adapt to climate change? Insights from Nepal's community forestry. Climate Dev 12(10):876–887. https://doi.org/10.1080/17565529.2019.1701971

Brooks N (2003) Vulnerability, risk and adaptation: a conceptual framework (38)

Caglar G (2013) Gender mainstreaming. Polit Gend 9(3):336–344. https://doi.org/10.1017/S1743923X13000214

Canelón SP, Boland MR (2020) A systematic literature review of factors affecting the timing of Menarche: the potential for climate change to impact women's health. Int J Environ Res Public Health 17(5):6–13. https://doi.org/10.3390/ijerph17051703

Carr ER, Thompson MC (2014) Gender and Climate Change Adaptation in Agrarian Settings: Current Thinking, New Directions, and Research Frontiers. Geogr Compass 8(3):182–197. https://doi.org/10.1111/gec3.12121

CCICED (2021a) Special policy report on global climate governance and China's role

CCICED (2021b) CCICED annual policy report 2020 green consensus and high quality development

CCICED (2021c) Report on gender mainstreaming in SPS research for the period 2020–2021c, Gastronomía ecuatoriana y turismo local

CEDAW (2009) 44th session 'Statement of the CEDAW Committee on Gender and Climate Change', (August)

Chen L (2014) Rural China's invisible women: a feminist political economy approach to food security 10(1):31–49

Chen L (2016) Gender statistics and local governance in China: state feminist versus feminist political economy approaches. China Q 225:190–213. https://doi.org/10.1017/S0305741015001630

Chindarkar N (2012) Gender and climate change-induced migration: proposing a framework for analysis. Environ Res Lett 7(2). https://doi.org/10.1088/1748-9326/7/2/025601

Daly M (2005) Gender mainstreaming in theory and practice. Soc Polit 12(3):433–450. https://doi.org/10.1093/sp/jxi023

Dankelman I et al (2008) Gender, climate change and human security lessons from Bangladesh, Ghana and Senegal. The Women's Environment and Development Organization (WEDO),

ABANTU for Development in Ghana, ActionAid Bangladesh and ENDA in Senegal, pp 1–73. https://doi.org/10.1016/j.neuroimage.2009.12.027
De Feyter K et al (2021) Gender equality: the formula to accelerate implementation of the 2030 agenda for sustainable development. In: Encyclopedia of law and development, pp 95–98. https://doi.org/10.4337/9781788117975.00030
De la Torre-Castro M et al (2017) Gender analysis for better coastal management—Increasing our understanding of social-ecological seascapes. Mar Policy 83:62–74. https://doi.org/10.1016/j.marpol.2017.05.015
Demetriades J, Esplen E (2008) The gender dimensions of poverty and climate change adaptation. IDS Bull 39(4):24–31. https://doi.org/10.1111/j.1759-5436.2008.tb00473.x
Denton F (2002) Gender and climate change : giving the "Latecomer" a head start
Denton F (2004) Gender and climate change: giving the "latecomer" a head start. IDS Bull 35(3):42–49. https://doi.org/10.1111/j.1759-5436.2004.tb00133.x
Djoudi H, Brockhaus M (2011) Is adaptation to climate change gender neutral? Lessons from communities dependent on livestock and forests in Northern Mali. Int For Rev 13(2):123–135. https://doi.org/10.1505/146554811797406606
Eastin J (2018) Climate change and gender equality in developing states. World Dev 107:289–305. https://doi.org/10.1016/j.worlddev.2018.02.021
Ergas C, York R (2012) Women's status and carbon dioxide emissions: a quantitative cross-national analysis. Soc Sci Res 41(4):965–976. https://doi.org/10.1016/j.ssresearch.2012.03.008
ESCAP (2018) Population dynamics, vulnerable groups and resilience to climate change and disasters:1–17. https://doi.org/10.1093/oxfordhb/9780199560103.003.0007
Federici S (2016) Women, reproduction and globalization. Économie Mondialisée Et Identités De Genre 2002:57–78. https://doi.org/10.4000/books.iheid.6171
Figueiredo P, Perkins PE (2013) Women and water management in times of climate change:participatory and inclusive processes. J Clean Prod 60:188–194. https://doi.org/10.1016/j.jclepro.2012.02.025
Gilley B (2012) Authoritarian environmentalism and China's response to climate change. Environ Polit 21(2):287–307. https://doi.org/10.1080/09644016.2012.651904
Ginige K, Amaratunga D, Haigh R (2009) Mainstreaming gender in disaster reduction: Why and how? Disaster Prevent Manag Int J 18(1):23–34. https://doi.org/10.1108/09653560910938510
Goh A (2012) A literature review of the gender—differentiated impacts of climate change in developing countries Structure. Capri 106:1–38. https://doi.org/10.2499/CAPRiWP106
Graziano K, Pollnac R, Christie P (2018) Wading past assumptions: gender dimensions of climate change adaptation in coastal communities of the Philippines. Ocean Coast Manag 162:24–33. https://doi.org/10.1016/j.ocecoaman.2018.01.029
Gumucio T, Rueda MT (2015) Influencing gender-inclusive climate change policies in Latin America. J Gend Agric Food Secur 1(2):42–61
He XJ (2017) Information on impacts of climate change and adaptation in China. J Environ Inf 29(2):110–121. https://doi.org/10.3808/jei.201700367
Heggelund G (2007) China's climate change policy: domestic and international developments. Asian Perspect 31(2):155–191
Hilft BE (2017) World risk report: analysis and prospects, pp 1–47. https://reliefweb.int/sites/reliefweb.int/files/resources/WRR_2017_E2.pdf
IPCC (2021) Policy Recommendations of the China Council for International Cooperation on Environment and Development. https://doi.org/10.1017/9781009157896.001
IPCC (2022) Climate Change 2022-Impacts, adaptation and vulnerability-summary for policymakers, Ipcc
Kesavan PC, Swaminathan MS (2006) Managing extreme natural disasters in coastal areas. Philos Trans R Soc Math Phys Eng Sci 364(1845):2191–2216. https://doi.org/10.1098/rsta.2006.1822
Klein LE (1995) American Society for Eighteenth-Century Studies (ASECS) Gender and the public/private distinction in the Eighteenth century: some questions about evidence and analytic

procedure published by : The Johns Hopkins University Press. Am Eighteenth-Century Stud 29(1):97–109
Kristjanson P et al (2017) Addressing gender in agricultural research for development in the face of a changing climate: where are we and where should we be going? Int J Agric Sustain 15(5):482–500. https://doi.org/10.1080/14735903.2017.1336411
Leung ASM (2003) Feminism in transition: Chinese culture, ideology and the development of the women's movement in China. Asia Pac J Manag 20(3):359–374. https://doi.org/10.1023/A:1024049516797
Lewis JL (2007) China's strategic priorities in international climate change negotiations. Wash Q 31(1):155–174. https://doi.org/10.1162/wash.2007.31.1.155
Lowther Ed (2012) EU climate change 'worse for women' claim row—BBC News. https://www.bbc.com/news/uk-politics-17774187 Accessed 19 Feb 2021
Luliu L, Hongmei X, Shiming M (2014) Understanding of climate change impacts: adaptation and vulnerability in urban areas and rural areas. https://doi.org/10.3969/j.issn.1673-1719.2014.04.005
Mcleod E et al (2018) Raising the voices of Pacific Island women to inform climate adaptation policies. Mar Policy 93(March):178–185. https://doi.org/10.1016/j.marpol.2018.03.011
Meier P, Celis A (2011) Sowing the seeds of its own failure: implementing the concept of gender mainstreaming. https://doi.org/10.1093/sp/jxr020.
Mersha AA, Van Laerhoven F (2016) A gender approach to understanding the differentiated impact of barriers to adaptation: responses to climate change in rural Ethiopia. Reg Environ Change 16(6):1701–1713. https://doi.org/10.1007/s10113-015-0921-z
Mikkola M (no date) No Title, Feminist perspectives on sex and gender
Moser C, Moser A (2005) Gender mainstreaming since Beijing: a review of success and limitations in international institutions. Gend Dev 13(2):11–22. https://doi.org/10.1080/13552070512331332283
NDRC (2007) China's National Climate Change Program
Nyasimi M et al (2018) Inclusion of gender in Africa's climate change policies and strategies. Clim Change Manag:171–185. https://doi.org/10.1007/978-3-319-69838-0_11
Oliver P (1991) What do girls know anyway?: rationality, gender and social control. Feminism Psychol:339–360. https://doi.org/10.1177/0959353591013003
Omolo N (2011) Gender and climate change-induced conflict in pastoral communities: case study of Turkana in northwestern Kenya. Afr J Confl Resolut 10(2):81–102. https://doi.org/10.4314/ajcr.v10i2.63312
Ongoro EB, Ogara W (2012) Impact of climate change and gender roles in community adaptation: a case study of pastoralists in Samburu East District, Kenya. Int J Biodivers Conserv 4(February):78–89. https://doi.org/10.5897/IJBC11.174
Onwutuebe CJ (2019) Patriarchy and women vulnerability to adverse climate change in Nigeria. SAGE Open 9(1). https://doi.org/10.1177/2158244019825914
Parpart JL (2014) Exploring the transformative potential of gender mainstreaming in international development institutions. J Int Dev 168(10–13):1–30. https://doi.org/10.1002/jid
Pearse R (2017) Gender and climate change. Wiley Interdisc Rev Climate Change 8(2):1–16. https://doi.org/10.1002/wcc.451
Peterson VS (2018) Problematic premises: positivism, modernism and masculinism in IPE. In: Handbook on the international political economy of gender, pp 23–36. https://doi.org/10.4337/9781783478842.00007
Piao S et al (2010) The impacts of climate change on water resources and agriculture in China. Nature 467(7311):43–51. https://doi.org/10.1038/nature09364
Porter F, Sweetman C (2005) Mainstreaming gender in development: a critical review. J Gend Stud 15(2):111. http://search.ebscohost.com/login.aspx?direct=true&db=ioh&AN=3336579&site=ehost-live
Preet R et al (2010) The gender perspective in climate change and global health. Glob Health Action 3(1):5720. https://doi.org/10.3402/gha.v3i0.5720

Reckien D et al (2017) Climate change, equity and the Sustainable Development Goals: an urban perspective. Environ Urban 29(1):159–182. https://doi.org/10.1177/0956247816677778
Rodenberg B (2009) Climate Change Adaptation from a gender perspective: a cross-cutting analysis of development-policy instruments, change
Salehi S et al (2015) Gender, responsible citizenship and global climate change. Women's Stud Int Forum 50:30–36. https://doi.org/10.1016/j.wsif.2015.02.015
Savo V, Morton C, Lepofsky D (2017) Impacts of climate change for coastal fishers and implications for fisheries. Fish Fish 18(5):877–889. https://doi.org/10.1111/faf.12212
Schwerdtner Máñez K, Poulsen B (2016) Fish is women's business too: looking at marine resource use through a gender lens. In: Perspectives on oceans past, pp 1–211. https://doi.org/10.1007/978-94-017-7496-3
Shabib D, Khan S (2014) Gender-sensitive adaptation policy-making in Bangladesh: status and ways forward for improved mainstreaming. Climate Dev 6(4):329–335. https://doi.org/10.1080/17565529.2014.951017
Sorensen C et al (2018) Climate change and women's health: Impacts and policy directions. PLoS Med 15(7):1–11. https://doi.org/10.1371/journal.pmed.1002603
Squires J (2005) Is mainstreaming transformative? Theorizing mainstreaming in the context of diversity and deliberation. Soc Polit 12(3):366–388. https://doi.org/10.1093/sp/jxi020
Su B et al (2018) Drought losses in China might double between the 1.5 and 2.0 °C warming. Proc Natl Acad Sci USA 115(42):10600–10605. https://doi.org/10.1073/pnas.1802129115
Su Y et al (2017) Gendered responses to drought in Yunnan Province, China. Mt Res Dev 37(1):24–34. https://doi.org/10.1659/MRD-JOURNAL-D-15-00041.1
Tandon N (2015) Empowerment of women in a green economy in the context of sustainable development and poverty eradication the case for community-based, gender-equitable and human rights-based gree. https://doi.org/10.13140/RG.2.1.1592.3368
Terry G (2009) No climate justice without gender justice: an overview of the issues. Gend Dev 17(1):5–18. https://doi.org/10.1080/13552070802696839
Trung PT (2014) Climate change and its impact on agriculture in vietnam. Int J Sci Res Publ 4(4):1–6
UN (1997) Report of the Economic and social council on gender mainstreaming, A/52/3(September):1–10. http://www.journals.cambridge.org/abstract_S1743923X13000214
UNWOMEN (2016) Gender dimensions of vulnerability to climate change in China, paper knowledge: toward a media history of documents
Wang W et al (2020) Barriers and requirements to climate change adaptation of mountainous rural communities in developing countries: the case of the eastern Qinghai-Tibetan Plateau of China. Land Use Policy 95:104354. https://doi.org/10.1016/j.landusepol.2019.104354
Watts N et al (2015) Lancet commission on climate change and health: health and climate change: policy responses to protect public health. Lancet (London, England). https://www.thelancet.com/journals/lancet/article/PIIS0140-6736(15)60854-6/fulltext
Wood HJ (2019) Gender inequality: the problem of harmful, patriarchal, traditional and cultural gender practices in the church. HTS Teol Stud/Theol Stud 75(1):1–8. https://doi.org/10.4102/hts.v75i1.5177
World Health Organization (2014) Gender, climate change and health. WHO Library Cataloguing-in-Publication Data. https://doi.org/10.17843/rpmesp.2016.331.2006
Yang F et al (2018) Acculturation of rural households participating in a clean development mechanism forest carbon sequestration program: a survey of Yi ethnic areas in Liangshan, China. J Econ 32:135–145. https://doi.org/10.1016/j.jfe.2018.04.006
Zhanga Y, Huanga L, Chaoa Q, Yangb Q (2021) Analysis of gender equality in climate governance. https://doi.org/10.1016/j.cjpre.2021.05.010
Zhao H et al (2015) Investigation and analysis about the adaptation on climate changes in rural area. Climate Change Res Lett 04(03):160–170. https://doi.org/10.12677/ccrl.2015.43018

Zhou M, Hu T (2021) Analysis of carbon emission status under the carbon neutral target in China for Earth's atmospheric balance. IOP Conf Ser Earth Environ Sci 804(4). https://doi.org/10.1088/1755-1315/804/4/042082.

Zhou Y, Sun X (2020) Toward gender sensitivity: women and climate change policies in China. Int Fem J Polit 22(1):127–149. https://doi.org/10.1080/14616742.2019.1687001

Zhu Z et al (2020) How feminization of forest management drives households' adoption of technologies: evidence from non-timber forest products operations in China. Forest Policy Econ 115(March):102154. https://doi.org/10.1016/j.forpol.2020.102154

Ting Wang is a Ph.D Candidate at Graduate school of Economics, Kyoto University. Ting is originally from China and earned a bachelor's degree in Administrative Management from Xiamen University in 2018. She came to Japan in 2018 and studied East Asia Sustainable Economic Development Studies at Kyoto University. After she graduated in 2020 with a master's in Economics, she continued her doctoral study in the same program. She is currently completing her doctoral research related to the linkage between climate change adaptation and gender issues, which is especially concerned with gender disparities under climate change disasters and how to achieve risk mitigation through policy instrument development.

Chapter 9
Beyond Western Dichotomies of Power: Life-Centered Development, Reciprocity and Co-creation Within Nature

Stéphanie Eileen Domptail and Milica Kočović De Santo

Abstract The chapter opens up key dichotomies characterizing the Western cartesian worldview and modernity to shed light on the consequences of an understanding of the world characterized by binaries and to open up the doors towards a new space of flourishing. We understand decolonization efforts in economic thinking and science as attempts to destabilize this binary world organization and to create a new space of possibilities, enabled by a worldview where power relations are distributed anew. We beginn by identifying and questioning the divides between who is central and who is peripheral, who is developed and who is developing, who is studying and who is being studied, how knowledge is created and accepted, who is controlling and who/what is being controlled. At the same time, we identify concepts lived among some communities that transcend these dichotomies, growing some early branches and roots cracking the colonial and modern Western edifice of thought and values. These include real existing practices such as *Buen Vivir*, life-centered development, self-governance in Yugoslavia (as endogenous knowledge), reciprocal relations with the non-human world, co-creation with nature, maintenance of the life-basis and reproduction of life, beauty, gender mainstreaming, positionality in science. These ideas and their practice help us redefine our models of thought and analysis, and contribute to the making of a worldview where our identity as humans may become newly defined.

S. E. Domptail (✉)
Institute for Agricultural Policy and Market Research, University of Giessen, Giessen, Germany
e-mail: Stephanie.domptail@agrar.uni-giessen.de

M. K. De Santo
Institute of Economic Sciences, Faculty of Dramatic Arts, Belgrade, Serbia
e-mail: milica.kocovic@ien.bg.ac.rs

M. K. De Santo and S. E. Domptail (eds.), *Degrowth Decolonization and Development*,
https://doi.org/10.1007/978-3-031-25945-6_9

9.1 Introduction

Alarming prospects for the future follow from the major crises threatening most people's well-being and life. They urge researchers and science to identify concepts and pathways for a transformation towards post-growth sustainability. Moore (2015) as well as Capra (1984) understand these multiple crises as a single crisis of consciousness, characterizing the hegemonic Western culture. Culture and imaginaries are crucial elements determining practices and behaviors, governance structures, as well as the production of knowledge recognized as valid (Sanford 2011; Hawkes 2004).

The Western secular Cartesian worldview is characterized by the concept of Modernity on the bright side, with its attributes progress, development and salvation, and by the concept of Colonialism on the dark side, with its attributes sexism, racism, extractivism and the invention of a Nature outside human beings. Western modernity is a narrative constructed in Europe and North America mainly, which has put in place institutions and languages that benefit those who built this imaginary (Escobar 1992). The Western worldview organizes the world in a binary, dichotomist manner reflecting uneven power relations. These dichotomies create a divide between who is central and who is peripheral, who is developed and who is developing, who is studying and who is being studied, who is controlling and who/what is being controlled. We understand decolonization in mainstream economic thinking and science as a critical attempt to destabilize this binary world organization and to create a new space of possibilities, enabled by a worldview where power relations are distributed anew. Degrowth is one form of such destabilization, compatible with decolonization processes.

Through a cultural approach, one looking at worldviews, we start deconstructing the Western hegemonic worldview on the nature-human relationship and on development, including growth. Our book *Decolonization, Degrowth and Development—Where culture meets the environment* addresses directly these central environmental, societal and political crises via a selection of case studies. The book relies greatly on ideas of political ontology to imagine a post-growth and post-development era (e.g. Escobar 1992, Sum and Jessop 2013, Castro 2004), questioning modernity and (neo-) colonialism, especially the colonization of nature and people. It asks and attempts to answer the questions: how does the Western worldview shape our economy and relation with(in) our environment? Why and how do people on the ground destabilize this hegemonic worldview practically and conceptually?

Our synopsis reflects on how practices and concepts described in the respective case studies counter or transcend specific dichotomies characteristic of the Western worldview. By doing so, this chapter provides food for thought to address very relevant and practical problems: How can we deal with the limited capacity of the economies to keep on growing in a finite natural world? How can we shape sustainable food and climate systems? How can we build a different relationship among ourselves and with(in) the Earth and our resources on Earth? Several practices and concepts strive to create and open a space beyond binary representations of the world, both in the domain of our relation to economic growth and of our relationship with(in) Nature. These concepts are real existing practices and include: *Buen Vivir*, self-governance in Yugoslavia (as endogenous knowledge), reciprocal relationships with the non-human world and co-creation with nature, maintenance of the life-basis and reproduction of life, beauty, gender mainstreaming and positionality in science. They constitute conceptual steps to decolonize our worldviews, and thereby our economy and our relation to nature.

In the following, we open up key dichotomies characterizing the Western Cartesian worldview and modernity to enlighten the consequences of an understanding of the world characterized by binaries and to open up the doors towards a new ontology for 'development', a new space of flourishing.

9.2 Questioning the Western Ontology: An Attempt to Transcend Western Dichotomies

Our means of transcending dichotomies are to raise awareness about the opposed concepts making up each dichotomy, as well as especially to show how these are culturally constructed and rooted in a political context conferring dominance to worldviews stemming from Western Europe and the politically dominant "West".

We define here the "West" as the cultural paradigm of the Modern, characterized by reductionism, imperialism, patriarchy, racism, and the Scientific Method (Mignolo and Walsh 2018; Merchant 2006). The "West" as a concept represents the cultural, economic and political entity mostly geographically located in Western Europe and North America, and extends to include other players ruling the Global North, such as (global) corporations, the influence of which grows in the (neo-) colonization processes and is eased up by acceptance processes of auto-colonization within Global South countries (Latouche 2015). Its power and values are contested both in the geographical West by marginalized actors, as well as in the rest of the world by populations feeling globally or locally marginalized and struggling to decolonize their economic, social and natural environment, that is the global South, the fourth World (indigenous politics) and other minorities. Their task is to decolonize imaginaries, by finding new symbols, values, meanings, flags, as alternatives to both the material and the mental dimensions of (neo-)colonization (Latouche 2015).

West *versus* East

Before the fall of the Berlin wall in 1989 and of the USSR, and as a consequence of this conceptualisation of the "West", eastern European countries were excluded from the "West" (and from the global North) and constituted the entity of the "East". Eastern European countries were governed in socialist structures with different types of socialism (egalitarian—centralized and self-governing—decentralized) ideology, while the West was driven by capitalism and a consumerist ideology. As shown by Pomieciński, by Kuźma and by Kočivić De Santo (this book), after the end of the cold war, the socialist countries transitioned abruptly to a capitalist economy and were suddenly confronted to the Western values and institutions through the work of global unification processes. Despite this tentative swallowing up, these regions now called post-transition and developing countries did not become fully 'Western' in the cultural sense, nor did they feel they now belonged to the global North. As Kuźma (Chap. 3) suggests, East European countries do not identify with the global South either. East European cultures remain in a space inexistent (or not visible and accepted in the manner of cultural pluralism) in the binary representation of the world into Global North and Global South. In parallel, or maybe as a result, as Kuźma reports, they also seek in these political interspaces or intervals what progress and development shall mean to them in the future.

From a global perspective, the dominance of Western on Eastern Europe materializes through the fact that "the West" refers mainly to Western Europe and not Eastern Europe. In fact, the concept of Eurocentrism refers to an excessive focus on Western Europe and not on the whole geographical Europe. Thus, the concept of the West with its western worldview and its Euro centrism masks the existence of a diverse Europe and grants a hegemonic character to Western Europe (Centre vs Periphery) in defining the European culture. Importantly, the Western hegemony of capitalism and consumerism with its neo-classical theoretical support in neo-liberal economics does not only dominate. Be it before or after the fall of the wall, it was the center of accepted thought and other ideas considered peripheral, were not granted attention or even were suppressed. In fact, Kočović De Santo (Chap. 2) attempts to restore attention to political and governance concepts that were *lost in the transition.* The cultural barrier was decisively hermetic to socialist thought possibly stemming from the East and a specific type of knowledge was systemically excluded. Even Central European knowledge created and published in the West, such as books and concepts of Czeslaw Milosz (the Captive Mind, Ketman, etc.), were only temporarily accepted to be subsequently excluded. Knowledge created in Western European countries, such as *animation socio-culturelle,* were also suppressed in Western Europe presumably because the terms did not fit the capitalist paradigm), while they were embraced in Brazil, Poland, Yugoslavia and other world semi-peripheries.

However, as demonstrated by Demuth (2019) in her remarkable and prized account of the environmental history of the Bering Strait, both East and West—the capitalist and the socialist systems—were characterized by materialism, a focus on production and production indicators dissociated from natural realities, that is, a reductionist and extractive worldview and economy. This common feature led to environmental disasters in all regions of the world, albeit in the name of different ideologies. In Europe, Kuźma as well as Kočović De Santo (Chaps. 2 and 3) show that during the 1970's both capitalist and socialist countries saw the emergence of counter movements appealing towards a reconnection with nature. A fundamental questioning of industrialization was expressed by some artists and intellectuals, for instance in Poland and Yugoslavia and later on in Armenia (Pomieciński, Chap. 4). This common appeal for a redefinition of the relationship with nature was and remains a common space upon which a space beyond East and West, and global North and global South can be built.

Center *versus* periphery, local *versus* global

The representation of a global North opposed to a global South overshadows the power relations and diversity in ontology and interests within countries of both North and South. Kočovic De Santo, Pomieciński, Richter and Domptail et al. (this book) show that throughout the world—in the global South (Ecuador), in developing regions (Armenia, Serbia) and in the global North (Germany) -groups of indigenous people, of farmers and of rural dwellers attempt to break the colonial relation to nature (controlling and extractive) and to well-being (consumerism or materialism), by developing and nurturing an alternative ontology, such as that of *BuenVivir* or agroecology, among other radical social movements. These cracks from within are what Mignolo and Walsh (2018) refer to in their depiction of decolonization. The government and transnational private corporations monopolize a political and economic sector and ontology where growth is the political goal, and around which radical social movements and opposition from local communities in rural areas gravitate. The latter feel exploited to the benefit of the center: they denounce environmental degradation and overexploitation (Pomieciński, Chap. 4). On the other side, they are perceived as a social and political challenge by actors of the 'center'. Their struggles are remarkable. Especially, governments and industries are rooted in the modernist worldview. They attempt to co-opt transformational concepts, such as agroecology for instance in the agricultural domain (Holt-Gimenez and Altieri 2013), by institutionalizing them in forms that do not question the world's order. Freeing oneself from this to some extent servile relationship to the centers and from an ontology which is embraced by most surrounding and resource controlling elements who attempt to hog transformational concepts, reveals very challenging (Domptail et al., Chap. 6) for actors of the periphery.

Growth *versus* Degrowth

The goal of Growth is an essential feature of modernism and of the Western paradigm. Although the growth paradigm was questioned neither in capitalist nor in socialist thought directly, the ideas of Degrowth and Decolonization permeated the socialist culture as well, even if the terms did not literally appear in their debates in today's form (Kuźma, Chap. 3). Reflections conducted by practical movements contributed to the factual de-colonization and independence in Global South countries and thereby to the Non-alignment Movement networking as shown by Kočović De Santo (Chap. 2). The Decolonization and Degrowth ideas became an inspiration to seek solutions and new visions outside the prevailing mainstream. They shattered the habits of thought of the societies they penetrated, exposing, for example, their inadequacy. They allowed the subversion to crack the monolith of, for example, one ideology or one authority. However, they were also subject to reinterpretation when colliding with other structures.

The creed of production, progress and growth as the goals of our human economies and lives drives the economic activity but also is intrinsically relates to our relation with(in) nature as shown by Richter and by Domptail et al. (Chaps. 5, and 6) among indigenous and agroecology communities. Questioning the very goal of growth thus requires the understanding that overcoming the human-nature dualism is an essential step towards reconciling social with ecological pressures. If Degrowth aims to address the civilization crisis, it must recognize the importance of culture. In other words, the Degrowth framework mustn't just enable people to think and act towards the halt of the ecological breakdown and social inequalities; it must address the prevalent anthropocentrism in our culture, societies, and economic systems.

Within corporations, the financial power mixes with a view of Nature as resources, which can and even should be exploited to create growth and well-being. The case of the Amulsar mine in Armenia (Pomieciński, Chap. 4) shows that opponents to the mine use Degrowth as a concept to rebel against the view of nature as a private resource fueling economic growth. In the public sphere, ideas of steady state economy and zero, non-growth or Degrowth, are seen as means of rebellion and resistance to the *status quo*. In addition, the opponents to the mine use Degrowth as a tool to decolonize this human-nature divide, to change the human relationship with(in) nature. Their discourse associates the protection of nature with the local culture, with a daily way of life and its survival. The argument rests on cultural democracy and basic human right(s) and creates a framework for economic and political decentralization and decolonization.

Finally, Richter (Chap. 5) shows that that the socioeconomic and ritualistic behaviours she observed in Ecuador maintain ecological equilibrium. Reciprocal relationships with the non-human world thereby constrain harmful growth and generate material and spiritual wellbeing within the human world. Domptail et al. (Chap. 6) also show that an affective or spiritual connection to nature can spark or even is intrinsic to changing economic goals from growth to the reproduction of societies and of life.

Progress *versus* backwardness

Progress is a major attribute of Western modernity (Mignolo and Walsh 2018). Opposed to progress in the neo-colonial dualism is the word backwardness. Backward qualifies all those who have not engaged in the race for progress through growth and materialism and applies to colonial subject and any other endogenous worldviews (understand local, indigenous, from within). In addition to the obvious extraction of people and nature as resources from colonized areas to the benefit of European actors of the New World and Europe, colonialism spread and applied the values of Western modernity to other cultures, structures and areas. Especially, Western values were seen as positive and superior. Within the 'Western colonial' worldview, the attributes 'wild and wicked' applied both to Nature, women and to colonial subjects (Merchant 2020), whose 'civilization' depended upon contact with and transformation by statutory legal, economic, social and cultural (Western) European standards (e.g. hunting permits and land registration for a recent example in Southern Africa, Domptail and Nuppenau 2018). In the last 70 years, growth was assumed to bring salvation to all "developing" nations. Thereby, the Western worldview sets standards defining the very identity and aims of human beings.

Alternative ideas of progress and humanity are inherent parts of the process of decolonization. As a first example, the decades of the second Yugoslavia (middle of 1950s to the end of 1980s) saw the rise of several initiatives testing alternatives to the "Iron Curtain" in Eastern Europe. According to Kočović De Santo (Chap. 2), Yugoslavia produced a dissonant endogenous know-how heritage in the domain of self-governance socialism developed within an internal project of social (cultural, economic etc.) organization. This knowledge was developed through experiences in few firms first, and then expanded gradually to other spheres of the economy and life, following people's decisions. This system was a unique socialist project oriented towards the market economy. By enabling all the political spheres to become more socially constructed, Yugoslavia introduced social ownership to achieve a decentralization of economic and political power. Secondly, at the international level, the role of the second Yugoslavia was essential in launching the non-aligned movement. It represented an instrumental alternative of time –by making visible and finally connecting the countries of the Third world. The humanity of the Third world was recognized. This movement encouraged factual decolonization and independence and enabled to start fresh international relations, in terms of cultural, educational and economic exchanges.

Today, the practitioners of concepts of Degrowth, of *Buen Vivir* (Richter, Chap. 5), of agroecology (Domptail et al., Chap. 6), of gender mainstreaming (Wang, Chap. 8), aesthetics of catastrophes (Schneider, Chap. 7) or proposed life-centered development based on previous practical and theoretical findings, as well as the critical voices emanating from developing countries in Eastern Europe (Kuźma, Kočović De Santo, Chaps. 2 and 3) and in Armenia (Pomieciński, Chap. 4) continue to rebel against the dominant hegemonic meanings of development, of civilization, of progress, of a feeding agriculture and its mirror backwardness. For instance, the life-centered development is presented by Kočović De Santo (Chap. 2) as an alternative to the current definition(s) of development. The idea of the life-centered development is that cultural, social and environmental spheres should be at the center of our future integrative management systems, because they create life and sense (for some authors—happiness). Thus, instead of mining social and environmental spheres to maximize our economic (capitalist) outputs, life-centered development recognizes that a society organised around principles of autonomy and direct democracy has the potential to meet the essential goals such as (by becoming nothing more than operational public policy) the social, cultural and the environmental aspects of our lives. These concepts are no less than an attempt to redefine "progress", "civilization", and thus, the human identity.

Human/culture *versus* nature, Anthropocentrism *versus* Ecocentrism

The dichotomy of Man versus Nature is deeply engrained in the Western worldview and at the origin of the environmental crises of the Anthropocene, as a major cosmological element ruling (hu)mans' relations with(in) nature. Indeed, cosmology, that is, dominant understandings of ontology, epistemology, temporality, cosmogony, and the role or place of humans in the universe, characteristic to a specific society in a given time and place (Allan 2018, p.11 in Richter 2022, p. 233), shapes our relationship with the non-human world. Some elements of the animistic cosmology have continued to rule people's management of nature despite the prevalence of

Christianism in pre-capitalist Europe. Nature was perceived as a living, nurturing female, and accordingly, treated in a way that allowed resources such as stocks of fish, game etc. to recover. Early merchant capitalism, internal enclosure in Europe and colonialism outside Europe ushered in a new era of social control, specifically over women, and an urge to control a now deemed wild, wicked and untamable nature (Merchant 2020, among many other same-minded eco-feminist authors).

The scientific revolution in Europe gave rise to the need, interest and ability to increase our control over nature through a detailed knowledge of it (Merchant 2006). This perspective is still present in the Western paradigm today as shown by Schneider (Chap. 7) through the analysis of artworks following the Katharina hurricane in the USA. According to her, the artwork demonstrates that the American artists who crafted the artworks share the belief that human beings are the most significant living entities on Earth. This worldview originates from the Christian principle of human dominion over Earth—*dominum terra.* As people in New Orleans lost control over the extreme climatic events, a feeling of powerlessness rose up and the demand to God to fix this. As opposed to this anthropocentric view (but maybe it is more a Men-centric view), Schneider shows how the ecocentrism in Asia (Japan) expresses a fully different relation of people with(in) nature. In line with Japan's primary religion, *Shintō*, the Japanese poet reported upon refers to the natural circle of life, and reveals a perspective placing intrinsic value on all living organisms and their natural environment. The inherent structure reveals an acceptance of life as it is, also awe towards nature.

In the midst of these two extremes, the experience of the agroecology farmers, striving to collaborate with Nature in Germany is remarkable. Domptail et al. (Chap. 6) show that farmers of la via Campesina- Germany fear and are in awe towards Nature. At the same time, they feel they design nature through their farming activities, and thereby walk a tightrope between control and *lasser-faire* of Nature. Yet, in the end, the farmers are creating a path for co-creation with Nature. Domptail et al. identify concepts indicative of a decolonized relationship with nature: *Design* ("*Gestaltung*"), *Co-creation* with nature, and the maintenance of the *life-basis* illustrate transformative possibilities that emerge from agroecological practices as a radical entanglement of humans with nature. In the same way, we can learn from the reciprocal human-nature interactions of those communities in Ecuador that put *Buen Vivir/sumak kawsay* into practice as introduced by Richter (Chap. 5).

Matter *versus* mind

The Western Cartesian worldview aims to explain the world without having to refer to the spirit, be it sorcery or religion. In that sense, it played a tremendous role in reducing the impact of the Christian institutions in the life and governance of people and their resources. At the same time, a vision of the world as a mass of organized matter whose behavior can be explained by only (bio)physical mechanisms negates the existence of a spiritual and emotional dimension in the human relation with(in) nature. In their respective contributions, Richter, Domptail et al. and Schneider (this book) show that across cultures (victims of hurricanes in Japan and the US, farmers in Germany, indigenous groups in Ecuador) emotional ties are key in their attitude and actions towards Nature. These include awe, fear, and desire to control, wonder, submission, humility, fulfillment, ease, fairness, empathy, and belonging. We argue that in the Western culture, the role of the affective dimension in the relation with(in) nature is taboo. Affective values cannot guide rational decision-making. Yet, agroecology farmers in the German case study experience the materiality of their interaction with Nature and experience a reunification of mind and matter. In addition, as the Cartesian Baconian worldview is reductionist in essence, it is interesting to note that both in Japan in the face of Hurricanes (Schneider, Chap. 7) and in Germany among agroecology farmers (Domptail et al., Chap. 6), a holistic vision of nature dominates: the complexity of the earth system and of nature at smaller scales is perceived as an element of inspiration, creativity, beauty and marvel. The theory of aesthetics presented by Schneider (Chap. 7) helps us to relativize our perception of and emotions towards Nature, as a cultural attribute. Thus, agroecology farmers just as indigenous communities in Ecuador entertain with much effort their own cultures and relations with(in) nature, despite their necessary embeddedness in the capitalist and market system to ensure their livelihoods.

Men *versus* women

The western worldview has established a patriarchal system, which originated with the "death of Nature" during the seventeenth century, as demonstrated by Merchant (2020) in her book. While monotheist as well as many other cultural systems were essentially patriarchal, modern capitalism developed as a process contingent to the further subordination and exploitation of women, nature and colonies. Patriarchy, as a structure, rewards masculine ways of thinking, behaving, and being with socio-economic privileges. It has evinced women from the public life to keep them grounded to the private household care and reproduction sectors. Yet, this economic domain—the *Oikonomia* of Aristotle—lost ground as mercantilism grew and chrematistic economics dominated the economic space. As a result, household management and economics is absent from formal economics. Women were associated with emotionality, subjectivity and thus irrationality; the exact opposites of the qualities of men and of the process of knowledge creation through the scientific method, which was seen as objective and rational. The scientific method of Bacon enabled men to tickle (or torture, Merchant 2006) out the true nature of Nature, including women, by looking at nature with a reductionist, detached and external perspective. The mechanistic (Capra 1984) perception of how nature works leads men to search for technical solutions to existing problems. This is still the case, as exemplified here by Wang (Chap. 8). In her analysis of policies for the mitigation of the climate crisis in China: only scientific and technical solutions are pursued by the government.

Wang (Chap. 8) further shows that in China, the patriarchal structures and their role in the climate change crisis are not questioned by the policies in place. Mitigation policies do not even refer to gendered effects of the crisis and to gender-sensitive measures to counter the stronger vulnerability of women. Yet, there is evidence that women appear more vulnerable than men to the multiple crises, including climate change, because of the patriarchal structures (Mies and Shiva 1993). At the same time, their voices often are ignored in decision-making. The feminist perspective on the ecological crisis and development (Merchant 2006) supports the cultural decolonization of development and modernity through the realization that women lack decision-making power because they remain invisible. In China (as an example), women remain hidden by the absence of differentiation between genders in management policies (Wang, Chap. 8).

In other settings though, as demonstrated by Richter and Domptail et al., farmers in the case studies established an affective relation to nature and their animals (Germany), or even a spiritual one (Ecuador) with a greater feminine Nature. In Ecuador (Richter, Chap. 5), the worldview regarding nature as a living female entity is alive, and can be said to lead to the establishment of ecological equilibria, as opposed to scenarios characterized by artificial scarcity of 'natural resources'. In Germany (Domptail et al., Chap. 6), the farmers revealed that their aim is the maintenance of a life-basis, both for themselves as for future generations, and other people and species on the globe. They apply the goal of reproducing themselves and the life-basis for human and other species in their decision-making. This attempt to concomitantly address their own survival and contribute to a life-basis for all is a tightrope exercise for them, especially in the context of an exploitative attitude towards Nature in Western modernism. Both groups of farmers thus re-introduce in the farming activity at least, in their relation to Nature, in their decision-making process, values and reproduction aims that are considered feminine. These constitute hopeful cracks in the prevailing patriarchal representations and structures, and a step towards the decolonization of the human relation with(in) nature.

Scientific *versus* non-scientific knowledge

From the seventeenth century, science and the creation of (valid) knowledge through the scientific method were to provide salvation from the mystic and from sorcery. Science aimed at taming people and nature in order to improve the (male Western) human condition (Merchant 2020). The role of science in validating knowledge goes nearly unchallenged today. Science-based policy-making is called for today, as a solution to avoid favoring one or another party in decision-making. The image of an absolute science precludes its questioning and the fact that science too, is socially constructed. In this way, several pieces of knowledge can be ignored in the name of (absolute) science (the new God).

Fricker's explanation on epistemic injustices provides clear and welcome concepts concerning the equity of knowledge forms and "knowledge capitalism". Fricker (2007) defined two types of injustices: testimonial and hermeneutic. Testimonial injustice is the expression of academic imperialism and takes place when certain sources of knowledge are considered as inferior and not valid. Testimonial injustice created an invisible but pervasive class system within academia and between academia and other forms of knowledge. For instance, the knowledge of pre-capitalistic societies in the management of resources as commons has been ignored in the last four centuries until a few decades ago. This prejudice is often directed to scholars from the periphery (and other minority groups), or/and stemming from other knowledge forms (tacit knowledge e.g. of farmers, Polanyi 1983) or oral testimonies

(Janvid 2021), or archives from global south countries (Mbembe 2015). This is why one German agroecology farmer insists so vehemently that the knowledge created on his farm is neither esoteric nor random; it is proven with methods from the Scientific Method (Domptail et al., Chap. 6). Testimonial injustice is also how Yugoslav theory of cultural policy and knowhow on worker-led governance systems (Kočović De Santo, Chap. 2) was to a great extent ignored or lost as a dissonant heritage, as an epistemological erasure (Dragićević Šešić 2018). Methods for the exclusion of other worldviews can also be economic, as shown by Richter (Chap. 5): the agroecological women's association suffered a boycott by the local market when first trying to sell their organic produce. The semi-periphery and Global South knowledge, or other means of knowing (Anthony-Stevens and Matsaw 2020) are treated as less relevant to the global knowledge body, policy or decision-making.

Hermeneutic injustice concerns the content of the knowledge itself, rather than the carrier of the message (or in addition to it, as often the two forms of injustices are concomitant). It takes place when certain groups, such as any marginalized community, women's groups, alternative farmers' group, (…), are unable to express and articulate their knowledge and give it meaning from the perspective of the dominant group (Fricker 2007). In essence, these groups share other languages of valuation (Martinez-Alier 2002), characterized by different ontologies, value and narrative systems. The experience and the knowledge of these groups often fall into a conceptual gap: they cannot be explained or translated through the concepts and approaches produced by the dominant groups and academia. This struggle to even dare to give value to non-monetised elements of the farming system, for instance, among German agroecology farmers shows the difficulty for them to both be accepted as farmers by the farming community and to dare think in other terms (e.g. connectedness with nature, animals) as well (Domptail et al., Chap. 6). The same applies to the Amulsar mine protesters in Armenia (Pomieciński, Chap. 4) and to some indigenous communities in Ecuador (Richter, Chap. 5) who struggle to articulate a multiplicity of values while their non-monetary character deprives these values from validity in national and international policy making under the Western capitalist paradigm, especially since the neoliberal logic of growth, success and competitiveness gained predominance from the 1990's. Indeed, with the remarkable exception of the contribution of Kothari et al. (2019): Pluriverse, which pays tribute to pluriverse practices and concepts in economics, we do not know of economics textbooks exploring endogenous (except folkloric) knowledge. This section darkens the focus of international organizations and foreign policies on an unidirectional knowledge transfer.

Traditional concepts of a good life such as *Buen Vivir* reflecting some elements of indigenous worldviews (Richter, Chap. 5) or new collaborative attitudes towards nature, even co-creation with nature strived by agroecology farmers in Europe, are developed in the real life and not considered knowledge until science—including economics—recognizes these. The lessons learned from socialist self-governance in Yugoslavia (Kočović De Santo, Chap. 2) or by artists and environmentalists of early counter movements in Poland (Kuźma, Chap. 3) are essential for the future shaping of the degrowth policies, since they hold factual knowledge about socialist market economy, and on decentralization of power relations and decolonization. Even in terms of plural values, culture in the Yugoslav system had de facto an important role as a non-economic and arguably a less monetary sector: its contribution to the social and common needs was acknowledged as its social responsibility. The cultural sector at the time pushed strongly the shaping of the society and its developmental goals (Kočović De Santo, Chap. 2).

The analysis of Schneider (Chap. 7) shows that art through aestheticizing catastrophes can also express powerful messages, reflecting but maybe also influencing cultural, societal and environmental values. When art pieces about human relations to natural catastrophes are compared, cultural values become explicit and their legitimacy or desirability can be debated.

One way forward is thus to give a voice to sources of non-science knowledge. Yet of course, because we are writing a scientific book, this creates internal struggles for us researchers, as we will see below.

9.3 Science's New Clothes? Decolonizing Scientific Analysis

Our hope is that our book operates as a decolonizing "fact" or "tool", and not only as a report on the decolonization of development. Thus, we want to reflect here on the frame of collective scientific publication. We too walk the tightrope of wanting to abide by the standards of scientific publication in our disciplines, while at the same time attempting to remain sensitive to and communicate about the cultural differences of interpretations that necessarily exist. The walk is made even more difficult by the fact that we use terms accompanied by long traditions of thoughts and consequent political resonances, such as degrowth and decolonization.

One first important achievement of this book is to gather authors who are, in their collective as well as individually, interdisciplinary and breaking the boundaries of disciplines in science. In fact, we see individual disciplinary integration as instrumental in raising our personal awareness to the relativity of disciplinary norms and standards in science. Our interdisciplinary background enables us to address environmental problems from a cultural and social perspective. In addition, as a collective, through the lens of cultural studies as our common denominator, we bring together political science, political economy, political ecology, economic history, decolonization studies, anthropology, art sciences, feminist economics, climate governance, environmental sociology and agricultural sciences to question the bright and dark faces of the Western worldview: modernity and colonialism.

In terms of content, our different cultural backgrounds in Western and Eastern Europe, Asia, as well as our attempts as Europeans to report on American or Asian worldviews, all feed into the critique of modernity and reveal perspectives that individually, we did not understand. This highlights the necessity to realize how our own culture limits our understanding of social realities and the need to work across culture, when it comes to addressing global crises and dethrone the Western worldview from its hegemonic position to land it as one of pluriverse co-existing ontologies.

We do not only contest objectivity in research, we contest the separability of researcher and researchees. The dichotomy subject vs. object is central to the scientific method, developed by Bacon, among others, in the seventeenth century and represents a heritage in modern science (Capra 1984; Merchant 2006). While researchers are expected to be rational, distant and objective, social science has long acknowledged that our prior knowledge baggage inevitably interferes with our objectivity when receiving, measuring/reporting and analyzing data. According to Anthony-Stevens and Matsaw (2020) or Basile et al. (2018), not only our personal baggage, but also our suitcase of tools, of methods and of approaches both in the way we collect and analyze our data on the one side and the theories we use on the other, are contingent on the specific scientific culture they emanated from, most times the Eurocentric Western worldview and its Scientific Method. Thus, in investigating ways in which groups of people attempt to decolonize their worldview, to radically change their action and human identity, we researchers are just as much as the "subjects of the research" constrained by the paradigm we grow in. Further, the interaction between the "subjects" and the researcher, that is, often, between the interviewee and the interviewer, is itself conditioned by this paradigm as shown in the contributions of Richter and Domptail et al. (Chaps. 5 and 6). Interviewees adapt their answers consciously to their perceived expectations of interviewers. To keep face, they may also unconsciously tend to choose arguments granting them status in the worldview, which dominates their lives, and thus showing rationality and logic, in the Western ontological context. Accepting these interferences as the norm is accepting that we as researchers have to communicate about our positionality—in situated knowledge(s) (Haraway, 1988).

In this book, we attempt to be transparent about the authors' beliefs and of those we write on behalf of. That was especially difficult when our concepts for analysis reflect the Western worldview. For instance, when analyzing German farmers' worldviews (Domptail et al., Chap. 6), the aim was not to categorize farmers' worldviews ultimately as colonial or decolonial, thereby constructing decoloniality as a normative target that can be acquired linearly through sufficient decolonization and the elimination of differences (cf. Mignolo and Walsh 2018: 81). Rather, we attempted to explore and qualify farmers' worldviews in their complexity and nuances. Our perspective aims at a process of healing integration, in the sense of a return or rediscovery of concealed and silenced elements of an original whole beyond entities of humans and nature, which have been forcibly removed over the past five centuries (Merchant 2006). In her contribution, Richter (Chap. 5) as a European white woman investigates the cosmology of Central America peoples, and faces a similar challenge. Due to the situated and contingent nature of any of such interviews, Richter reported tensions between her own cognitive-political spatial temporality (that of scientific-empiric, libertarian-emancipatory enquiry that itself is grounded in European philosophy and political thought) and that of the interviewee (Scott 1995). The tension arose regardless of whether the intended outcome was to challenge and transcend ways of thinking that are based on a dichotomous separation between knowing and being in the first place (Wood 2020).

Emancipatory research as put forward by this volume attempts to dismantle colonial structures of domination. This is antithetical to the dominant, "extractive" process of knowledge creation between researcher and researchee. To avoid appropriating indigenous knowledge, Richter engaged with indigenous scholarship prior to, during and after the fieldwork process. She interviewed indigenous activists and academics as well as their allies who are involved in the construction of *Buen Vivir*, not as a development paradigm co-opted by the state, but as a decolonial grassroots practice that is contested and transformed by every-day practices.

In this volume, we consciously and deliberately work against "reproducing colonialist moves that produce discourse under conditions of unequal power" (Escobar 1995, p. 9), to the limits of our means. The reproduction of colonialist discourses would further homogenize and essentialse the participants of our research and would allow us to exercise the kind of discursive power this volume is attempting to dismantle (ibid; Mohanty 1988). The book also attempts to be transparent in the identity and background of the authors to make our position that knowledge is also socially constructed evident. This is meant as a provocation in a scientific context where generalisability (beyond cases and also beyond analysts) is the key to validity.

9.4 Conclusions

The chapter questions and discusses major dichotomies anchored in the so-called Western secular Cartesian worldview characterizing European Modernism, and its accompanying coloniality. These dichotomies construct the political, the economic, the epistemological spheres of this ontology, as well as the relation of humans with(in) nature and range from Global North/South, West/East, Centre/Periphery, global/local, growth/ post-growth, mind/matter, humans/nature and object/subject. Several concepts providing alternatives to this binary thinking emerged from the discussion. These concepts and historical lessons are listed in Table 9.1: self-governance and *BuenVivir* towards life-centred development, affective abundance, co-creation with nature, maintenance of the life-basis and reproduction of life, decolonization of knowledge creation. These concepts enlighten existing endogenous (non)Western knowledge, beyond anthropocentrism, gender mainstreaming etc.). Not only do these concepts transcend the binary thinking and help us redefine our models of thought and analysis, they also contribute to the making of a worldview where our identity as humans is newly defined. In this process, the people and groups actively engaged in this ontological rebellion, upon which our book reports, adapt their consciousness to the new global environmental, social and political context of crises humanity faces. They contribute some early branches and roots cracking the colonial and modern Western edifice of thoughts and values.

How to operate cultural change remains a key question, which this book can only slightly touch upon. We have seen through the collection of works that a diversity of worldviews co-exists de facto on our planet. Yet, they have unequal power in their influence on the management of resources and the economy, as well as on individual

Table 9.1 Western Dichotomies and concepts decolonizing the Western worldview (*Source* authors)

Dichotomies characterising the „Western" worldview	**Concepts contributed by or used in *Degrowth, Decolonisation and Development* transcending these dichotomies**
Global North versus Global South West versus East	**Pluriverse** (Kothari et al. 2019) **Degrowth as a unifying concept between West and East** (mentioned in Kuźma)
Center versus periphery Local versus global	**Rural–urban linkages** with "little centres of the World" (mentioned in Pomieciński) **Glocalisation** (Krzysztof Czyzsewski, Sejni, Czeslaw Milosz) **Non-aligned cosmopolitanism** (Milena Dragićević Šešić 2018)
Growth versus non-growth	**Degrowth** (D'Alisa 2019, D'Alisa et al. 2020, Hickel and Kallis 2019, Jackson and Victor 2020, Martinez-Alier et al. 2014) **Frugal abundance** (Latouche 2015) **Ecological equilibria** (mentioned in Richter) **Spiritual and Material wellbeing** (mentioned in Richter)
Progress versus backwardness	**Epistemic injustices** (Fricker 2007)
Human (culture) versus nature, Anthropocentrism versus Ecocentrism	**Co-creation with(in) Nature** (Domptail et al.) **Life-centred development** (Kočović De Santo) **Transformation of the inside** (versus changing the outside) (Schneider) **Humans in the cycle of life** (Schneider) **Reciprocity with the non-Human world** (Richter)
Matter versus mind	**Frugal abundance** (Latouche 2015) **Dialogue with nature** (Domptail et al.) **Spirituality** (mentioned in Domptail et al.) **Spiritual and Material wellbeing** (mentioned in Richter)
Men versus women	**Reproduction of the life-basis** (Domptail et al.) **Gender mainstreaming** (mentioned in Wang)
Scientific versus non-scientific knowledge	**Holism** (Lovelock 1970s) **Equity of knowledge** (Ezeanya-Esiobu 2019) **Inter-epistmic dialogue** (Richter)
Subject versus object	**Positionality in science** (Haraway 1988)

behaviors. The Western worldview dominates in the domains of law, governance, the economy, ethics, science- all key domains of human and international exchange, as well as interactions with(in) nature. Even at the local level, as the case studies in this book report, deciding, working and valuing Nature outside the anthropocentric extractive worldview is a struggle against others' and even one's own cultural inheritance as central Europeans. That is how cultural change from within becomes a difficult tightrope exercise.

Culture is also reflected by our institutions. Chang (2007) remarks that economic conditions shape our way of thinking, acting, and relating to the world. As much as our economic system influences how we view the world, for instance, in the way in which capitalist development has led us to regard nature as a resource, "*culture [also] changes with economic development*" (Ibid, p. 182, emphasis added). For instance, there are evidences on neoliberalism shaping consumer subjectivities, how economic interests of merchant capitalism in Europe aided in the construction of the mechanism metaphor (Merchant 2020). If culture shapes institutions and institutions shape culture, then how to break this positive reinforcing feedback of the Western paradigm remains the question for further analysis that shall rely on the historical evidences on transformative role of culture.

In the process, science can contribute more than to reveal values upon which to build an alternative ontology. Science, as man-made activity, rooted in the modernist Western worldview, shall transcend the boundary between subject and object. All researchers are rooted and conditioned by their own cultural belief system and in that of the scientific method. We never can see the world as a bat (Nagel 1974). Making our (cultural, power) position transparent is one way to legitimizes us to make such radical demands for a cultural change. They must also question in a constructivist fashion the roots of the methods and theories (especially their basic assumptions) modern disciplines are founded upon.

The westernization through the colonial processes excluded many other alternatives, by aligning the knowledge on how development shall look like. In order to support global economy of capitalism, capital seeks cheaper resources – both human and natural. The real existing and often endogenous knowledge show that there are many possible alternatives and pluriverse lenses existing in parallel. The growing civil disobedience and resistance in response to extractivist policies is a clear call for collective action for a different political and economic course and in a needed transition. We firmly believe that we ought to find quick and steady solutions for post-development and post-growth futures.

Acknowledgements We express our special gratitude to all authors of this book, who contributed to this chapter by highlighting their key results and conclusions. The chapter has greatly benefited from reviews and supporting comments from Martin Petrick, Milena Dragićević Šešić, and Katharina Richter. Stéphanie Eileen Domptail's work with Jennifer Hirsch, Jessica Mörsdorf and Sergej Enns also inspired this chapter.

References

Anthony-Stevens V, Matsaw SL Jr (2020) The productive uncertainty of indigenous and decolonizing methodologies in the preparation of interdisciplinary STEM researchers. Cult Stud Sci Educ 15(2):595–613. https://doi.org/10.1007/s11422-019-09942-x

Basile S, Asselin H, Martin T (2018) Co-construction of a data collection tool: a case study with atikamekw women 17(3):840–860. https://acme-journal.org/index.php/acme/article/view/1414.
Capra F (1984) The turning point: science, society, and the rising culture. Bantam Books, New York
Castro CJ (2004) Sustainable development: mainstream and critical perspectives. Organ Environ 17(2):195–225. https://doi.org/10.1177/1086026604264910
Chan H (2007) Bad samaritans. the myth of free trade the secret history of capitalism. London, Bloomsbury Press
D'Alisa G, Demaria F, Kallis G (2020) Degrowth: słownik nowej ery. LangeL, Łódź
D'Alisa G (2019) The state of degrowth. In: Chertkovskaya E, Paulsson A, Barca S (eds) Towards a political economy of degrowth. Illustrated edition. Rowman & Littlefield Publishers, London, pp 243–57
Demuth B (2019) Floating coast: an environmental history of the bering strait. W.W. Norton, New York, p 416
Domptail SE, Nuppenau EA (2018) Push and pull relations between villages and cities from a rural perspective. Rural 21 Int J Rural Develop 52(4)
Dragićević Šešić M (2018) Culture and democracy: cultural policies at work; Is cultural policy theory possible at the semi-periphery (Kultura i demokratija—kulturne politike na delu—da li jemoguća teorija kulturne politike na poluperiferiji?), journal Kultura, 2018, n. 160, pp 13–35, link to English version, https://www.casopiskultura.rs/wp-content/uploads/2021/09/prevodi/6.%20Milena%20Dragicevic%20Sesic,%20CULTURE%20AND%20DEMOCRACY,%20Culture%20No.%20160,%202018.pdf
Escobar A (1992) Imagining a post-development era? critical thought development and social movements. Soc Text 31(32):20. https://doi.org/10.2307/466217
Escobar A (1995) Encountering development: the making and unmaking of the third world. Princeton Studies in Culture/Power/History. Princeton University Press, Princeton, NJ
Ezeanya-Esiobu C (2019) Indigenous knowledge and education in Africa: Springer Open (Frontiers in African Business Research)
Fricker M (2007) Epistemic injustice: power and the ethics of knowing. Oxford University Press
Haraway D (1988) Situated knowledges: the science question in feminism and the privilege of partial perspective. Feminist Stud 14(3):575–599
Hawkes (2004) The fourth pillar of sustainability: culture's essential role in public planning. Cultural Development Network: Common Ground, Melbourne
Hickel J, Kallis G (2019) Is green growth possible? New Polit Econ:1–18. https://doi.org/10.1080/13563467.2019.1598964
Holt-Giménez E, Altieri MA (2013) Agroecology, food sovereignty, and the new green revolution. Agroecol Sustain Food Syst 37(1):90–102. https://doi.org/10.1080/10440046.2012.716388
Jackson T, Victor PR (2020) The transition to a sustainable prosperity-a stock-flow-consistent ecological macroeconomic model for Canada. Ecol Econ 177(November):106787. https://doi.org/10.1016/j.ecolecon.2020.106787
Janvid M (2021): Testimony in African epistemology revisited. S Afr J Philos 40(3):279–289. https://doi.org/10.1080/02580136.2021.1954766
Kothari A, Salleh A, Escobar A, Demaria F, Acosta A (2019) Pluriverse: a post-development dictionary. https://www.academia.edu/39692614/_Pluriverse_A_Post_Development_Dictionary_AUF_2019_NEW_BOOK_edited_by_Ashish_Kothari_Ariel_Salleh_Arturo_Escobar_Federico_Demaria_and_Alberto_Acosta_Download_full_ebook_for_free_PDF_License_Creative_Commons.
Latouche S (2015) Decolonization of imaginary. In: D'Alisa G, Demaria F, Kallis G (eds) Degrowth: a vocabulary for a new era. Abingdon, Routledge, 117–120
Martinez-Alier J (2002) The environmentalism of the poor. In: a study of ecological conflicts and valuation, pp 312. Edward Edgar Publishers. ISBN: 9781840649093. eISBN: 9781843765486. https://doi.org/10.4337/9781843765486

Martinez-Alier J, Anguelovski I, Bond P, Del Bene D, Demaria F, Gerber JF, Greyl L et al (2014) Between activism and science: grassroots concepts for sustainability coined by environmental justice organizations. J Polit Ecol 21(1) https://doi.org/10.2458/v21i1.21124.

Mbembe A (2015) Decolonising knowledge and the question of the archive. Basis for public lectures held in South Africa. Published by the University of Witwatersrand, South Africa under https://wiser.wits.ac.za/system/files/Achille%20Mbembe%20-%20Decolonizing%20Knowledge%20and%20the%20Question%20of%20the%20Archive.pdf. Accessed last on 2 Feb 2023

Merchant C (2006) The scientific revolution and the death of nature. Isis 97:513–533

Merchant C (2020) The death of nature: women, ecology, and the scientific revolution. HarperOne; Reprint Edition (10. Januar 1990)

Mies M, Shiva V (1993, 2014) Ecofeminism. Zed Books (Critique Influence Change), London, New York

Mignolo WD, Walsh CE (2018) On decoloniality: concepts analytics praxis. Duke University Press, Durham/London

Mohanty CT (1988) Under western eyes: feminist scholarship and colonial discourses. Fem Rev 30:61–88. https://doi.org/10.2307/1395054

Moore JW (2015) Capitalism in the Web of Life: ecology and the accumulation of capital. London/New York: Verso

Nagel T (1974) What is it like to be a bat? Philos Rev 83(4):435–450

Polanyi M (1983) The tacit dimension. Gloucester, MASS. Library of congress, Catalog card number 66-21015. First published in 1966. ISBN: 0-8446-5999-1

Richter K (2022) Provincialising degrowth and situating buen vivir: a decolonial framework for the politics of degrowth (Doctoral Dissertation). Goldsmiths, London

Sanford AW (2011) Ethics, narrative, and agriculture: transforming agricultural practice through ecological imagination. J Agric Environ Ethics 24(3):283–303

Scott D (1995) Revolution/theory/modernity notes on the cognitive-political crisis of our time. social and economic studies 44(2/3):1–23. https://www.jstor.org/stable/27866025

Sum NL, Jessop B (2013) Towards a cultural political economy. Edward Elgar Publishing. https://doi.org/10.4337/9780857930712

Wood DA (2020) Epistemic decolonization: a critical investigation into the anticolonial politics of knowledge. Springer Nature

Stéphanie Eileen Domptail is a senior researcher at the professorship for Agricultural, Food and Environmental Policy at the Justus Lieibig University of Giessen, Germany. Stéphanie applies concepts of ecological economics and feminist economics to the study of agricultural and food systems. She investigates agroecological systems in Germany, Nigeria and Brazil. She brings together the topics of ethics, decolonization and agroecology in their joint capacity to redesign human-nature relationships towards a more desirable state.

Milica Kočović De Santo is an interdisciplinary senior researcher at the Institute of Economic Sciences in Belgrade, Serbia. She investigates alternative systemic solutions in the quest for needed radical transitions through social theories, involving economic, developmental, decolonial, commons, cultural studies which gather in the field of degrowth. Milica also teaches at the Faculty of Dramatic Arts and University of Arts in Belgrade, Serbia at MA and Ph.D study levels.

The manufacturer's authorised representative in the EU is Springer Nature Customer Service Centre GmbH, Europaplatz 3, 69115 Heidelberg, Germany. If you have any concerns regarding our products, please contact ProductSafety@springernature.com

Printed and bound by CPI Group (UK) Ltd, Croydon, CR0 4YY
28/09/2026
02224071-0002